George Ogilvie

The Genetic Cycle in Organic Nature

Or, the succession of forms in the propagation of plants and animals

George Ogilvie

The Genetic Cycle in Organic Nature
Or, the succession of forms in the propagation of plants and animals

ISBN/EAN: 9783337231682

Printed in Europe, USA, Canada, Australia, Japan

Cover: Foto ©berggeist007 / pixelio.de

More available books at **www.hansebooks.com**

THE

GENETIC CYCLE IN ORGANIC NATURE;

OR, THE SUCCESSION OF FORMS IN THE PROPAGATION OF PLANTS AND ANIMALS.

BY GEORGE OGILVIE, M.D.,

REGIUS PROFESSOR OF THE INSTITUTES OF MEDICINE IN THE UNIVERSITY OF ABERDEEN; AUTHOR OF THE "MASTER BUILDER'S PLAN IN THE TYPICAL FORMS OF ANIMALS."

ABERDEEN: A. BROWN & CO.
EDINBURGH: JOHN MENZIES.
LONDON: LONGMAN & CO.

MDCCCLXI.

ABERDEEN:
PRINTED BY D. CHALMERS AND COMPANY,
ADELPHI COURT, UNION STREET.

PREFACE.

The objects of the present work are, firstly, to determine the mutual relations of those diversified forms which the recent researches of Naturalists have shown to be propagated from each other, in many different species among the lower orders of both kingdoms of Nature; and, secondly, to consider what analogy can be traced between the successive forms in these species, and the phenomena which occur in the reproduction of those higher in the scale of organization, where, to all appearance, like always produces like.

The Author is well aware that a much abler pen than his would have but little chance of enlisting the interest of any large circle of readers in such a work as the present; for, though it would not be easy to point to any subject in the whole range of Biological Science more calculated to excite feelings of wonder and enquiry, yet the obscurity of some of the questions involved, while it has stimulated the zeal of many earnest investigators, has no doubt kept back others from giving the same attention to this, as to other branches of Physiology; besides which, it is not to be denied that the Natural Sciences—rapid as has been their advance of late in popular favour—are still far from occupying their fair place in public estimation. A conviction of this would probably have prevented his ever setting of his own accord to write for publication on the subject. But this work was written, in the first instance, for a different purpose, having been undertaken originally in response to a call made with reference to the late Meeting of the British Association at Aberdeen, though, as it came to be too unwieldy for this, an abstract was afterwards substituted, which was subsequently printed in the *Edinburgh Philosophical Journal* for January of last year.

The primary object of bringing forward the communication was to elicit discussion on views for the systematizing of one of the most perplexed questions in Physiology, and the remarks to which it gave rise, though necessarily restricted, went to invite their fuller examination. For this end, the Author is now led to publish the Treatise itself, in all essential points as it originally stood, though he has submitted it to a careful revision in matters of detail, and has re-written a large part, for greater clearness of expression.

At the same time, he must admit, that when his work had taken shape, he felt—as in the case of a former one on the Unity of Organization—that it went, in so far, whether fitly or not, to occupy a void in biological literature. As the nature of the subject, however, precludes its consideration, except by *bona fide* students of Natural Science, he has made no attempt, in the present work, to treat it in a popular way, his object having been simply to make use of such expressions as seemed best fitted to convey his meaning. If, as he has some reason to fear, his anxiety for this, and his wish to present the subject in different points of view, have led him into needless repetitions, he can only now express his regret that circumstances latterly have not allowed him leisure to determine what might be retrenched with advantage.

In explanation of the absence in the text of any reference to the accompanying Plates, it is but fair to mention that their introduction was an after-thought. The difficulties in the way of publication interfered with the illustration of the text by wood-cuts, as is now commonly done in treatises of this kind, and it was not till after the work was in print that the use of outline figures was suggested. Much selection was then impossible, but the Author hopes that those which have been introduced will be found an advantage, from the acknowledged importance of illustrative figures of some kind, when frequent reference has to be made to peculiarities of structure and conformation.

To personal researches on the subject in hand the Author cannot make much pretension; he can claim little more than to have taken for many years a lively interest in the observations of others. He must, therefore, readily admit his liability to fallacies and misconception in making his statements at second hand, though there may be some compensation for this in a comparative freedom from influences which are apt to bias the views of original observers. He may at least, in all honesty, claim to have entered on the consideration of the question without any conscious prepossession one way or other; and, although the conclusions here brought forward have for some time back fully approved themselves to his judgment, he is free to admit that his views had previously undergone many and important modifications since this department of Physiology first became the subject of his particular attention. But in the present state of the Science, indeed, no conclusions can well claim more than a provisional acceptance.

It only remains to acknowledge the kind assistance received, in various ways, from several friends, and particularly from Professors Sharpey, Simpson, and Allen Thomson, and to express the hope that the work may contribute something, at least by placing facts in a new point of view, to the elucidation of a much-perplexed department of Natural Science.

CONTENTS.

I.—DERIVATION OF ORGANIC BEINGS.

                        Page.

1.—Parental Derivation a distinctive character of Organic Bodies, ...   1
2.—Grounds for disallowing the theory of a Physical Derivation or Spontaneous Generation of Living Beings, ...   2
3.—Derivation from one, or from two Parents, .   9
4.—Tendency of these modes of propagation to Alternate, or recur periodically, when they co-exist in the same species, ...   10
5.—Distinction of Cases of Alternation into *protomorphic*, *orthomorphic*, and *gamomorphic*, according to their relations in the Genetic Cycle, . .   12
6.—Indications of Representative Phenomena in the Higher Species, ...   15

II.—SURVEY OF THE REPRODUCTIVE PROCESS IN THE VEGETABLE KINGDOM.

1.—Introductory Remarks, ...   17
2.—General character of the process in the Vegetable Kingdom, ...   20
3.—Reproduction in the Protophyta, . . .   22
4.—In Algæ generally, and Characeæ, . .   32
5.—In Fungi and Lichens, ...   40
6.—In Hepaticæ and Mosses, . .   49
7.—In Ferns and Equisetaceæ, .   52
8.—In Lycopodiaceæ and Rhizocarpeæ, .   55
9.—In Gymnospermous Phanerogamia, .   57
10.—In Angiospermous Phanerogamia, .   61
11.—Concluding Remarks, . . .   66

III.—Survey of the Reproductive Process in the Animal Kingdom.

		Page.
1.—General Character of the Process,	. .	68
2.—Reproduction in the Protozoa,	. . .	74
3.—In the Cœlenterata,	. . . .	81
4.—In the Echinodermata,	. . .	84
5.—In the Polyzoa,	. . . . .	87
6.—In the Tunicata,	. . . .	89
7.—In the higher Mollusca,	. . .	91
8.—In the Helmintha,	. . . .	94
9.—In the Annelida,	. . . .	99
10.—In the higher Articulata,	. .	102
11.—In the Vertebrata,	. . . .	106
12.—Concluding Remarks,	. . . .	107

IV.—Nature and Varieties of Alternation of Generations.

1.—Points of Agreement and Diversity in cases of Alternation, ... 109
2.—*Protomorphic* Alternation, as in the Trematoda, from Gemmation in the Early Stage of Development, ... 112
3.—*Gamomorphic* Alternation, as in the Polypifera, from a later Gemmation, in the evolution of Sex, 117
4.—Relations of these two forms of Alternation, . 128
5.—Their Co-existence in the case of the Cestoid Worms, 131
6.—*Orthomorphic* Alternation, as in the *Aphides*, from Gemmation in the evolution of the Typical Organization, ... 133
7.—Reference of the best known cases of Alternation, to one or other of the foregoing heads, . 136
8.—Co-existence of all the three forms in the case of some Annelida, ... 137
9.—Alternation in the *Salpæ*, ... 140
10.—Protomorphic and Gamomorphic Alternation in Mosses and Ferns among Plants, . . 142

V.—Pullulation.

1.—Notice of a Series of Successive Gemmations, interpolated at certain points, . . . 146

		Page.
2.—Most frequent in the Middle Stage of Development,		147
3.—Ramified disposition of Plants and Zoophytes from Adhesion of the Gemmæ,		149
4.—Such adhesion a point of minor importance in the present enquiry,		150
5.—Pullulation itself of secondary importance to Alternation,		153

VI.—EMBRYOGENY, AS REPRESENTING ONE FORM OF ALTERNATION.

1.—Representative Phenomena in the Higher Species,	156
2.—Representation in Embryonic Development,	157
3.—Illustration from Duplex Monstrosity,	157
4.—From the Development of the Polyzoa,	160
5.—From that of Cestoid Worms,	161
6.—From that of the Echinodermata,	162
7.—Essential Community of Nature,	164

VII.—REPRESENTATION OF THE OTHER FORMS.

1.—Representation of Orthomorphic Gemmation in the Higher Species,	166
2.—Representation of Gamomorphic Alternation in the Maturation of Sex,	167
3.—Illustration from the Periodic and Late Development of the Organs,	167
4.—From the case of the Polyzoa,	170
5.—From some aberrant Crustacea,	172
6.—From the later phases of the Cestoid Entozoa,	173
7.—From the reproduction of the Polypifera,	173
8.—Points of distinction between Organs and Zooids,	174
9.—Illustration from the reproduction of Phanerogamic and Cryptogamic Plants,	180
10.—Correspondence inferred between the process of Maturation and Gamomorphic Alternation,	188

VIII.—RELATIONS OF OVA AND GEMMÆ.

1.—Points of distinction between Ova and Gemmæ, with notice of transitional forms,	190

		Page
2.—Nature of the Germs in the Viviparous *Aphides* and allied Insects,	. . . . .	191
3.—Nature of the Ephippial and Common Eggs in Entomostraca and Rotifera,	. .	193
4.—Development of Unimpregnated Ova in certain Insects,	. . . . .	195
5.—Occasional development of Unimpregnated Ovules in Plants.	. . . . .	199
6.—Nature of the Viviparous Flowers of Plants,	.	200
7.—Indications of a tendency to development in the Sexual Elements singly,	. . .	202
8.—Essential community of nature between Ova and Gemmæ,	. . . . .	204
9.—Question of the necessary recurrence of Sexual Generation,	. . . .	206
10.—Bearing on the theory of Alternation of Generations,		208

IX.—SUMMARY OF CONCLUSIONS.

1.—Recapitulation of Results, - . . .	212
2.—Formula of the Genetic Cycle, . . .	216
3.—Modification for the higher Animals, . .	216
4.—Extension for Pullulation, . . .	218
5.—Concluding Remarks, - . . .	219

X.—CASES SIMULATING ALTERNATION OF GENERATIONS.

1.—Casual Alternation of Gemmation and Generation,	221
2.—Alternation of Diverse Gemmations, . .	221
3.—Atavism, ...	222
4.—Metamorphosis, ...	223
5.—Spermatophores and Sporophores, . . .	225

XI.—HOMOLOGICAL RELATIONS OF THE STRUCTURES CONCERNED IN THE GENETIC CYCLE.

1.—Nature of the ultimate Sexual Elements, .	229
2.—Homological Relations of the Germinal and Spermatic Elements, ...	234
3.—Links of the Genetic Cycle in the several Groups of Organized Beings, . - . .	237

		Page.
4.—Relations of the Sex to the Individual,		241
5.—Concluding Remarks,	. .	248

APPENDIX.

		Page.
TABLE I.—Of the Genetic Cycle in Plants,	.	254
II.—The Genetic Cycle in Animals,	. .	255
III.—Periods of Interpolation of Gemmation in different groups of Organized Beings,	.	256
IV.—Resting periods in the Genetic Cycle,	.	257
V.—Abstract of Dr. Sanderson's Table of Analogies in the Development of different Classes of Plants,	. . .	258
VI.—⎱ Analogies in the Genetic Cycle of Plants, VIII.—⎰ as indicated in this Work,	. .	261
IX.—⎰ Tabular Views of the Genetic Cycle in XVIII.—⎱ the different Classes of Animals,	.	266

ON THE GENETIC CYCLE.

I.

DERIVATION OF ORGANIC BEINGS.

ALTHOUGH there is no distinction in Nature more clear or more universally recognized than that between Organic and Inorganic Bodies, yet, when we descend to the lowest forms of the former, we find the marks which characterize them as a class become at last so little appreciable, that there is perhaps only one among those generally brought forward as diagnostic, which may be looked for as universally present —their *derivation*, by a process more or less direct, from previously existing individuals of a like kind.

In a certain sense, indeed, derivation from like forms may have place also in the origination of various substances simply physical in their nature, but—to pass over other differences—there is this obvious distinction, that it is not *essential* to their formation. When it has place, it may facilitate their production more or less, as we find the crystallization of a saline solution accelerated by the presence in it of crystals of the salt, already formed, which serve as nuclei for additional deposits of the like kind ; but such production will occur whenever the requisite chemical, mechanical, and other physical agencies come to operate upon matter in which the ultimate elements of the substances in question are present.

But with Organic Bodies it is different; for in their case not only must their ultimate chemical elements be present in some shape or other, but they must be present as combined by the prior operation of the living powers of individuals of a like kind into fertilized germs or other reproductive bodies. If such a germ or reproductive body has been normally constituted, then, and then only, will the application of certain appropriate influences of the nature of light, heat, chemical action, &c., become the means of its being developed into a body eventually resembling that from which it was itself derived.

§ 2. On the validity of this point of distinction between organic and inorganic forms, naturalists and physiologists, after some vacillation, are probably now pretty well agreed; the most recent researches affording as strong evidence against the origination of organic beings *de novo*, as any of a negative kind can well be. It is well known that organic matter in the state of decomposition into which it passes so readily when exposed to air and moisture at a suitable temperature, is found to swarm with minute forms of animal and vegetable life. But it has been shown by repeated, and apparently conclusive experiments, that no such development of living beings will occur if all the materials concerned in the experiment—that is, both the air and the organic matter, with water and other adjuncts—be subjected to processes which effectually destroy all particles they may contain of the nature of eggs or seeds, endued with a latent capacity of vital action; and this, though care be taken not to alter their nature so as to unfit them in any way for the *support* of life once developed.

The experiments of Professor Schultze, of Berlin, are among the first of the kind which were performed with the precision necessary in operations of this nature. They consisted in passing the air which was allowed to come in contact with the decomposing matter through a fluid (oil of vitriol), capable of destroying organic matter by actual contact, without emitting any noxious fumes. Though the air was constantly renewed, no production of living forms took place in the substances under observation, so long as the current of air was subjected to this filtration ; while in comparative experiments, in which the decomposing matter was freely exposed to the atmosphere, without the employment of any such sifting process, the usual development occurred of the lower forms both of animal and vegetable life.* Without being at all aware of these experiments, I made some myself, of a similar nature, and with the same result, by passing the air through red-hot capillary tubes. These I reported in a communication to the Parisian Medical Society, in 1843. Some years after, on the reports current of the discoveries of the late Mr. Crosse, the electrician, I repeated the experiments of Professor Schultze, with some modifications,—as by passing a continuous electrical current of low intensity through the organic matter,—but I have obtained always the same negative result. M. Milne Edwards also refers to experiments of his own, of the same general nature, and, like Schultze's, entirely opposed to the spontaneous development of organic forms.†

Quite lately, however, we have the report of experiments by two independent observers, affording a different result. I refer to those of M. Pouchet, communicated last year to the French Academy ; and those of Dr. Daubeny, which were brought under the notice of the British Association at its late meeting at Oxford. M. Pouchet's experiments consisted in macerating in distilled water a portion of the contents of a flask of hay which had been exposed (dry) to a high temperature in an oven for half-an-hour. The ap-

* For a detailed notice of these experiments, see the Edinburgh Philosophical Journal for July, 1837 ; also, Owen's Compar. Anat. I., 32.
† Edinburgh Philosophical Journal, Oct., 1859.

paratus was immediately sealed hermetically ; but, notwithstanding these precautions, Infusoria were soon developed in the contained fluid. To the cogency of these experiments, Milne Edwards takes objection on the following grounds :—

1. The inadequacy of the means employed to ensure the heating of the whole mass of hay to the boiling point, or to any temperature inconsistent with the retention of vitality.

2. The capacity of animals, such as the Rotifera—much higher in the scale than the simpler Infusoria—of recovering their vitality on being moistened after desiccation, and even after exposure in the *dry* state, to a degree of heat which would be fatal in the natural condition.*

Dr. Daubeny's experiments were of the same general nature as those of Schultze. No development of animal life took place ; but, notwithstanding all the precautions employed, mouldy vegetations made their appearance in the fluid. In the discussion which ensued on the reading of Dr. Daubeny's paper, two possible sources of fallacy were suggested—the employment of lint-seed meal luting, and the passage of the air through the oil of vitriol in bubbles of too large a size to ensure the full action of the caustic on all the suspended particles of an organic nature. But, as far as concerns the mucedinous Fungi, which give origin to these vegetations, it would appear that sulphuric acid, however employed, is no real barrier ; for M. Pasteur has found that they will bear even the prolonged contact of the concentrated acid without losing their power of germination.†

The experiments of M. Pasteur, in some other points connected with this subject, are so satisfactory in their results, both positive and negative, that a short reference to them is essential to complete our notice of the question. M. Pasteur first satisfied himself of the actual existence of

---

* Edinburgh Philosophical Journal, Oct., 1859.
† Annales des Sciences Naturelles. Ser. IV., tom. XII. (zool), p. 86.

organic particles in the air, by drawing it through a small dossil of gun-cotton, which was then dissolved in a mixture of alcohol and ether, and the residuary matters, after washing, examined by the microscope. Some of the particles thus obtained were evidently of an organic nature, as indicated by their form and structure. A good many appeared to be minute grains of starch, and were at once dissolved by concentrated sulphuric acid; but some corresponded with the spores of the mucedinous Fungi in their powers of resistance to that reagent. M. Pasteur then made various experiments to test the possibility of spontaneous generation, of the same general nature as those already noticed. In one of these a fermentible saccharine fluid was placed in a large glass flask, with the neck drawn out into a capillary tube. After prolonged boiling, the flask was allowed to fill itself with air in cooling, the capillary neck through which the air entered being made red hot, and afterwards hermetically sealed, when refrigeration was complete. No development of life took place so long as the flask was intact, though kept at a warm temperature for a month or even six weeks; but when a small plug of gun-cotton was introduced—charged in the way described with atmospheric dust — though precautions were taken which appeared abundantly sufficient to prevent the entrance of any other extraneous matters, a turbidity soon became apparent in the fluid, which was quite limpid before, and vegetations began to develop themselves in about the same time that they appeared in comparative experiments on liquids freely exposed to the atmosphere, gradually spreading from the vicinity of the ball of cotton through the whole contents of the flask.

These and other experiments of M. Pasteur's, bearing on the same subject, appear fully to justify his conclusion, that there is nothing in the air, save these dust-like germs, which can be the cause of the development of animal and vegetable life, following on its free access. Even the oxygen comes into play only as a supporter of the life of the forms originating from these germs. Neither gas nor liquid, neither electricity nor magnetism, nor ozone, nor anything else, known or unknown, which may be present in the air, save only the germs which it carries, are the essential condition of the development of life.*

The general result of experiments of this kind, taken in conjunction with arguments from the general analogy of plants and animals, have now led to the abandonment, by common consent, of the theories, once so prevalent among physiologists, of spontaneous generation ; for the clear inference from these experiments is, that there exist constantly, either in the organic matter, or, more probably, in the natural air or water, multitudes of germs of many different organisms, and that the speciality of the forms that appear in particular cases depends, either on the nidus variously modifying the development of germs originally identical, or on its so favouring the growth of some, that the others are stifled, as it were, and so prove abortive.

We have many other indications of the existence of multitudes of such germs floating in the air. It is mainly by the conveyance of pollen in this way that we account for the fertilization of the seeds of diœcious plants ; and that the spires of various species of ferns must be similarly wafted about in their vicinity, appears from the difficulty of obtaining pure seed to be depended on for the multiplication of any particular species, many of the plants raised turning out of quite a different kind from that from which the spores were collected. †

It is true that there are still many cases in which Naturalists are much perplexed to account for the formation of organized beings, if the idea of their origination indepen-

* Op. Cit. pp. 85-89.
† Dr. Balfour, in Edin. N. Phil. Journ. VIII., 278.

dently of parents—or their *spontaneous generation*, as it has been called—is to be held inadmissible. This is especially the case in regard to the development of parasitic beings ; but so much has been done of late years in the investigation of these obscure cases, as to render it highly probable that those as yet unexplained are no real exception to the general law of parental derivation. The facts which have come to light in the pursuit of such enquiries are of a kind equally surprising and interesting, but it would be out of place to refer to them more in detail here, as many of the phenomena will again come under notice in the farther treatment of the subject proposed for discussion in this work—the laws regulating the derivation of living beings from each other. It may suffice, therefore, to close this allusion with the following summary of the subject, in the words of Prof. Owen :—"The 'thread-worms' *(Filariæ)* of certain insects, which present no trace of sexual organs, were supposed to be spontaneously developed in those insects. The little worms were, however, by special and due research seen to wind their way out of the caterpillars they infested. Von Siebold placed these free *Filariæ* in damp earth, into which they soon bored : in a few weeks he found that the sexual organs were developed in them, and that they laid hundreds of eggs. Early in spring the young worms were hatched, and began to creep about. Von Siebold took some young caterpillars of the moth *(Iponomeuta evonymella)* in which were no parasites : he placed them in the soft earth in which the young *Filariæ* had been hatched, and, in twenty-four hours, most of the caterpillars were infested by the young thread-worms, which had bored their way through the soft skin, into the interior of the young caterpillars. The long hair-worm of fresh waters *(Gordius aquaticus)*, vulgarly conceived to be the result of a metamorphosis of the hair of a horses' tail, passes its early life as a parasite in the body of an insect. But many Entozoa acquire their full or sexual development, not as free worms, but within the body of another animal, and of a species distinct from that in which they had passed the early stage of their existence.*

In illustration, Professor Owen gives a summary of the reproduction of the Distomata, which we shall presently have occasion to consider more at length.

He proceeds :—" The sum of the recent researches on the generation of the Entozoa, teaches that, to the success in life of the majority of these internal parasites, two different species of much higher organized animals are subservient ; and that these two species stand in the relation of prey and devourer. The habits of the prey favour the accidental introduction—as when a slug crawls over the droppings of a thrush—of the eggs of the bird's intestinal parasite. These are hatched in the slug. The slug, in its turn, is devoured by the thrush, but the parasitic passengers are not digested—only the coach is dissolved, and the larvæ, thus set free, find in the warm intestines of the bird the appropriate conditions for their metamorphosis and full development. In like manner, the *Rhynchobothria* of a cuttle fish are the larvæ of the *Tetrarhynchus* or four-tentacled tape worm of a dog-fish. The encysted sexless *Triænophorus* of the liver of the char becomes the free and perfect *Triænophorus* of the gut of the pike. The *Ligula* of a herring becomes a *Tænia* only when introduced into the interior of a cormorant. The bladder-worm *(Cysticercus fasciolaris)* of the mouse's liver becomes the tape-worm *(Tænia crassicollis)* of the cat. The *Cysticercus pisiformis* of the liver of the hare becomes the *Tænia serrata* of the dog and fox. Dr. Kuchenmeister of Zittau first proved, experimentally, by feeding animals with *Cysticerci* (Hydatids of the flesh and glands of herbivorous animals) that they

---

* Address to Brit. Assoc. at Leeds, 1858. P. 23.

became *Tæniæ* (intestinal tape-worms) in carnivorous animals."*

§ 3. But, although it would appear that we may safely enough admit the universal derivation of living beings more or less directly from others of the same kind, which stand to them in the relation of parents, it is by no means so clear that we are entitled to assume any absolute uniformity in the way in which this law is carried into operation. In fact, even a superficial survey of nature must make us aware of one notable point of difference ; for, while in all the higher forms, we find two parents (or their representatives) concerned in the act of reproduction, we meet with many cases among those lower in the scale of organization in which a single individual appears capable of procreation by its own unaided powers. It will suffice at present to cite in illustration the case of the *Aphides* among insects.

The existence of these two different modes of origin—by single and double derivation — is now so universally admitted that special terms are in use for their designation —such as *homogenesis* or *monogenesis* for the former, and *heterogenesis* or *digenesis* for the latter.† The term *gemmation* (budding) is also used by many authors to denote propagation by single derivation, as distinguished from that higher form of generation which involves the combination of two original elements. In the former mode of origin a portion of the body of the parent becomes the seat of a

---

* Address to Brit. Assoc., 1858. P. 34.

† This is the sense in which the terms *monogenesis* and *digenesis* are proposed by Prof. A. Thomson (Cyclop. Anat. and Physical, Art. *Ovum*, Suppl. p. 42), and the sense in which they will be employed in the following pages; but it is necessary to observe that Prof. Van Beneden, in his extensive works on the reproduction of the Entozoa, uses them with a very different meaning; by *monogenesis* he understands direct development; and by *digenesis* the interpolation of intermediate forms, in the way of alternation—*i.e.* not as here, *genesis* from two origins, but *genesis* in two stages.

certain independent manifestation of vitality, and a focus of such intensity of the plastic processes, that, in the course of time, the part is converted into a distinct organism, capable of detachment from the parent, and fitted to maintain a separate existence. Such a detached gemma may be termed a *free zooid* or *phytoid*.

In the ordinary form of reproduction again—that by the co-operation of the sexes—a fusion seems to take place of two highly vitalized portions of the same or kindred organisms, which results in the formation of a *fecundated germ*, possessed henceforth of an independent vitality, endowed with a capacity for ultimately acquiring the structure characteristic of the species, and destined to be thrown on its own resources, by its extrusion from the protecting envelopes, as soon as its organization is sufficiently advanced for this condition. In all but the very lowest forms of life—the conjugating Algæ—a difference is observable between the two factors of the embryonic product, which are recognized respectively as male and female, or as the *spermatic* and *germinal* elements.

§ 4. It has long been known that these two modes of propagation may co-exist in the same species—the plant or animal multiplying now in one way, now in the other ; but the general relations of the two modes of increase to each other, throughout series of organised beings as a whole, has only recently engaged the attention of naturalists. Till lately, the general opinion seemed to be that the latter form of reproduction was the normal one in the higher species both of plants and animals, while derivation from a single parent prevailed in the lower grades of both kingdoms, there being some in an intermediate position, which furnished examples of the co-existence of the two forms. This arrangement was considered to be quite exceptional among animals, though of common occurrence, under some modifications, in the vegetable kingdom ; but no farther relation was indicated between the two processes in such cases, than that a high development of the one was usually accompanied by a proportionate abeyance of the other.

Now, in so far there can be no doubt that it is only in the lower forms that gemmation is met with as an obvious phenomenon; but the idea can no longer be entertained that sexual reproduction is confined in the same way to the higher species, the tendency of recent investigations being rather in favour of greatly extending the limits of both forms. In particular, an origin from two parents is now known to have place in many species, in which it had long been overlooked on account of its recurring only at intervals, the ordinary mode of propagation being by offsets from a single stock. Thus, in some of the lower Algæ, the ordinary mode of increase is by the formation of new cells, which, becoming detached from the parent frond, may form each the nucleus of a new and independent plant. This goes on with more or less vigour during the whole season; but when such a change of external circumstances supervenes, as to interfere with continuous growth, a new mode of propagation is brought into play in the process termed conjugation, or the coalescence of two cells in separate fronds, so as to form a seed-like body, which, after lying dormant till the return of conditions favourable to vegetation, gives origin to a new frond, like one of those previously produced by detached cells. Here the process of digenesis —represented by the fusion of two cells—comes in only at intervals, to supplement, as it were, what is in this case the more usual one of monogenesis—represented by the detachment and germination of single cells.

In those of the lower species, in which both modes of propagation are well marked features, we find that they have a tendency to succeed each other in a regular order, with corresponding differences in the immediate progeny, to which the term *alternation of generations* has been applied—an expression which, though open to some objections, has now come into very general use.

The recurrence of sexual reproduction, in some form or other, has now been ascertained in so many species, ordinarily propagated in the way of gemmation, as strongly to suggest the probability of the occasional interposition of the other form being eventually detected, by close and long continued observation, even in those cases in which derivation from a single source is the only mode of increase yet known; and the impression is gaining ground among naturalists that the two are co-existent, and alternate with more or less regularity and distinctness in many of the lower Invertebrata, and probably still more universally in the vegetable kingdom. It is unnecessary at present to go into any argument on the point, as the sketch which will presently be given of the modifications of the function of reproduction in both kingdoms of nature will serve to show the grounds on which this view is founded.

§ 5. I believe, indeed, that it falls short of the truth, and that there are processes co-extensive with organic nature which represent, in some degree, the gemmation of the lower species, and alternate in the same way with the sexual act.

I must premise, however, that, even in cases where the phenomena of alternation come out most clearly, I cannot regard the processes as all of a parallel kind. They appear to me to fall into groups whose characteristic features depend principally on the period of the life-history of the species at which a process of gemmation is interpolated in the genetic cycle.

The gemmation sometimes occurs just before, and is, as it were, ancillary to sexual reproduction; sometimes it occurs after it, when it is subservient rather to the progress of development. In the former case, what may on the whole be considered as the most typical of the diverse forms be·

longing to the species, is still defective in having no proper organs of reproduction—a function which is vicariously performed by a set of gemmæ detached from it. The original stock is really neuter ; but true sexes appear in these buds, after they have been transformed by a process of development into isolated zooids or phytoids. They may then be considered as a highly individualized form of those organs which were wanting in the parent stock. Such organs constitute, at least, the essential part of their economy ; and although, along with them, there may be present also others, more or less fully developed, for discharging functions, such as alimentation and locomotion, required by their status as free zooids, yet their great office is reproduction, and this end effected, their life speedily comes to a close. In this they contrast strikingly with the stock from which they were derived ; for it is endowed with a much greater permanence of life, frequently detaching, during its period of vigour, many successive swarms of sexual zooids ; just as among the higher animals, the same parent may develope many successive broods of young.

On the other hand, when the budding process occurs in the course of development, the gemmæ are detached from the immediate product of impregnation while it is still in a rudimentary condition, comparable to the first stage in the evolution of the ovum of the higher animals. The germ-parent itself never attains to the full development of the species, but remains the whole term of its brief existence in a rudimentary state ; but the progeny, which it buds off, acquire, in due course, the typical form, or at least give origin, mediately or immediately, to others which do so.

For the better distinction of these varieties of alternation, and for the purpose of bringing out more clearly what I conceive to be their points of correspondence with phenomena occurring in the higher animals, I have found it convenient to divide the life history of an organic being into three stages, all of which come out prominently in one form or other of alternation, while, as I shall endeavour to show, they are covertly represented even in those species in which no phenomena of alternation are recognised. The first, or what I term the *protomorphic* stage, is that which intervenes between the fecundation of the germ and the first appearance of the characteristic or typical organization of the species; the second, or *orthomorphic*, that which corresponds to the development and full perfection of this organization; while the third, or *gamomorphic*, is that of the formation or maturation of those structures in which the spermatic and germinal elements are generated, in preparation for another act of fecundation, as the commencement of a new genetic cycle.

In one of the forms of alternation just noticed, the interpolation of gemmation takes place in the protomorphic stage,—that is, prior to that development by which the features most characteristic of the species are gradually evolved. The other form of alternation, just contrasted with it, is that in which the process of gemmation is interpolated in the gamomorphic stage—that is, after the general acquisition of the typical conformation of the species, and in connection with the development of the organs which form the sexual elements.

In the intermediate period of the life history of the species—that here termed *orthomorphic*—which intervenes between the appearance of the general typical character of the family and the maturation of sexual organs, gemmation, though, perhaps, a more frequent character than either in the incipient or terminal stages, rarely comes before us as a case of alternation of generations, in consequence of the gemmæ commonly remaining in adhesion to each other, so that their separate individuality is never clearly manifested, and the whole aggregation passes as a single plant or animal. This is especially the characteristic arrangement in the vegetable kingdom, and in the zoophytic forms of animal life. Where the gemmæ do become detached, however, the case may assume the aspect of a form of alternation, one of the most striking examples—that occurring in the propagation of the *Aphides*—being, as I believe, referable to this head.

In cases, therefore, where alternation is an obvious and well recognized occurrence, I would distinguish these three varieties :—

1. That in which the gemmation occurs in the protomorphic or germinal stage, prior to the appearance of the typical organization ;

2. That in which it occurs in the gamomorphic or later stage of the life history—that is, the connection with the maturation of the reproductive organs ; and

3. That in which it occurs in the orthomorphic or intermediate stage—that is, during the manifestation of a more fully developed condition of the typical organization, but prior to the maturation of the sexual organs.

§ 6. But, even in the higher forms of life, where we no longer find any obvious alternation, I believe that phenomena occur which are in some degree representative, and which admit of the same sort of classification. Thus, though the well-marked cases of alternation, due to the evolution of protomorphic zooids, are confined to a few of the lower orders, a certain *nisus*, or tendency in this direction,—a fresh start, as it were, in the course of germinal development—may be traced with more or less distinctness in all cases of embryogeny, as in all instances there is formed first a cellular germ-mass, from one point of which there is subsequently developed a new axis of embryonic growth. And as the appearance of the new centre of organization in the early germ may stand as representative of protomorphic alternation, so to the contrasted form marked by the production of sexual or gamomorphic zooids, we may trace a certain correspondence in the maturation of the reproductive organs.

The subject of these relations I propose to examine with more detail after the survey, just referred to, of the reproductive process in organic nature, but it seemed necessary to premise so much here, as I have made use, in its course, of forms of expression, to avoid circumlocution, which would hardly be intelligible without such explanation.

II.

SURVEY OF THE REPRODUCTIVE PROCESS IN THE VEGETABLE KINGDOM.

WHAT is true in a degree of all physical science, applies with especial force to Biology — namely, that the so-called "General Laws," being really nothing else than general expressions of facts, can only be determined when the individual facts have first been clearly ascertained in a great variety of instances, and then carefully and minutely compared for the discovery of their mutual relations. A partial selection of facts can serve only as the basis of conjecture, and nothing, probably, does more to damage the progress of science than the abuse of hypotheses, to give a false show of symmetry and completeness to empty systems, which are the product merely of the speculator's fancy, and have no real counterpart in nature. Our dislike to confess ignorance, and go through investigations which promise no immediate result, predispose us but too much to adopt such ready-made systems, which must fall to pieces, indeed, at last from their own rottenness, but which, for a time, do much mischief by stifling the ardour of research, or giving it such a false bias as to cause facts and relations to be overlooked in a manner which seems almost incredible after the delusion has once passed away.

Yet it is admitted that there is a use of hypothesis in science which is perfectly legitimate, and which, when employed with due tact and caution, has proved of most signal service in the hands of many of the most gifted students of nature in unlocking her secret recesses.

What is it but such a guarded use of hypothesis that gives modern chemistry a claim to rank as a philosophical pursuit, which could never be awarded to the laborious but desultory researches of the old alchymists. In the main, chemistry still is, and must long remain, an empirical science : the very fluctuations of its nomenclature show that its so-called theories are no more than hypotheses, still they have the important use of giving system to what would otherwise be a chaos of unconnected facts ; and there is undoubtedly a certain amount of truth in the relations thus indicated, though in any fully developed system this is as obviously eked out by conjectures. To those who have the acuteness to distinguish between them, the former furnishes important indications for the course of farther investigations; and these, when completed, serve to determine the truth of the conjectural superstructure.

But, in fact, the natural tendency of the human mind to generalize makes it impossible to prevent the introduction of hypothesis. Few persons can address themselves to the consideration of any extensive series of phenomena without attempting to trace certain relations of this kind among them, and the validity of their conclusions is entirely a question of degree. In proportion as the facts are numerous and accurately known, and as they are carefully and judiciously compared, the results arrived at will be entitled to rank as well ascertained general laws ; but far short of this conclusions *will* be drawn. With the average constitution of minds, it is impossible it should be otherwise ; and there is, therefore, probably more good likely to be done by insisting on the necessity of distinguishing the well-founded from the merely conjectural, than by entirely excluding the latter.

Though we may admit, therefore, as is the opinion of some well qualified to judge, that we have not yet in most branches of biological science a sufficient knowledge of actual facts to admit of our conclusions from them ranking much above the level of hypothesis, we still reasonably expect to derive some advantage from their guarded use, especially in methodizing the more intricate departments; and none, surely, stands more in need of being systematized than that involving the consideration of all those diversified and seemingly anomalous phenomena, which have of late been so abundantly accumulated in connection with the function of reproduction. Only we must see that no hypothesis be admitted which is not in consistency with known facts, and jealously resist all attempts to bend facts into conformity with hypothesis.

It is with this view that I now propose, before proceeding to theorize on the nature and relations of the reproductive function, to give a summary of what I believe to be the main facts of the process in the principal groups of both kingdoms of nature. But I may observe at the same time that the conclusions which were indicated at the close of the last chapter are only partly of the nature of hypotheses —partly I consider them as matters of fact. Thus I believe it to be a matter of fact that there is a very essential discrepancy in the so-called cases of alternation of generations, and that of a kind to divide them naturally into groups as above indicated. It is, on the other hand, a matter of theory—and it may be of mere conjecture—that one class is represented by the process of embryogeny, and the other by that of the maturation of the sexual organs in the higher species. It is at least a question that requires to be argued, and this cannot be done till the facts have been brought forward; but that the difference referred to does exist in cases of alternation, is only, I think, what will appear in the course of the summary from the very facts of the case.

It is no part of my intention to discuss here the whole history of Reproduction. I propose to confine myself to such points as have some distinct bearing on alternation, but these I will endeavour to state, as distinctly as I can, and as fully as is consistent with the brief limits to which I feel that I should confine myself in such a summary.

As Reproduction, like other functions, is performed in a less complex way in the more rudimentary forms of organization, I propose proceeding from the lower to the higher species, and for similar reasons, I commence with the Vegetable kingdom.

§ 2. A few remarks may be made, by way of introduction, concerning the general characters of the process in this division of organized nature.

Indications of sexual relations in the function of reproduction, have now been so generally ascertained, as to make it highly probable that they are always present. Three principal modifications have been recognized in cases where their existence has been positively ascertained.

The most rudimentary manifestation of sexual action occurs in the conjugation of the lower Algæ. In these simple forms of living matter there is at first no differentiation of organs for the work of reproduction, more than for any other vital function. The plant consists simply of a repetition of cells, which—whether isolated or aggregated into fronds, (tumid, membranous, or filamentous)—are indistinguishable from each other, being all nucleated and filled with a green granular mass, termed endochrome. In the farther development of the cells, we find that in those which are to take part in the reproductive process, the endochrome becomes condensed into a spheroidal mass; after this, a communication is established between two neighbouring cells by a perforation of their walls, and a fusion of their contents takes place, resulting in the formation of an embryonic corpuscule, which eventually becomes a true cell, by the formation of a proper wall on its exterior. Along with a general sameness in the essential features of this process, in all the simpler Algæ or *Protophyta*, there are certain specific differences, mainly dependent on variations in the mode of perforation of the cell-wall, and the investment of the embryonic corpuscule.

In the higher Cryptogamia another modification of the reproductive process prevails, in which the sexual elements are represented by phytozoa and archegonial corpuscules. Linnæus was led to apply to the class in which he placed these species, the name of Cryptogamia, from his not being able to extend to it those principles of sexual relation which he recognised in the higher plants. But since his time, after much groping on wrong tracks, relations of this nature have at last been detected, more closely analogous in some respects to the reproductive phenomena of animals, than even those with which we were previously acquainted in phanerogamic plants. The analogy lies mainly between the spermatic or fertilizing elements in the two kingdoms, the vegetable phytozoa, being like the spermatozoa of animals, minute rounded or filiform bodies, endowed in general with motile powers by the action of ciliary appendages. The germinal element, which is lodged in the centre of a flask-shaped group of cells, termed the archegonium, appears to be at first a homogeneous mass of protoplasm, but after fertilization by the phytozoa, it becomes a true cell by the development of a distinct wall on its exterior. There is a great similarity throughout the group, in the general character of both kinds of corpuscules, and in their immediate investments, but very great diversity in their relations to the parent stock. Sometimes, as in mosses, they are developed at once on the leafy axis; in other cases, as in ferns, they are formed in detached phytoids, resembling distinct plants of a lower type—peculiarities which will come again to be noticed, in reference to the contrast between these two families, as representing opposite forms of alternation of generations.

The remaining modification of the reproductive process —that by pollen grains and ovules—characterises all the higher or phanerogamic plants, which constitute the vast majority of the present flora of the world. The ovule is a cellular body containing the germ, and developed from a peculiarly modified leaf or carpel, which in the Coniferæ leaves it quite exposed, but in by far the greater number of plants is wrapped round it to form the germen or ovary. The pollen grains are secondary cells developed in an anther or terminal appendage of a stamen—representing the lamina of a peculiarly modified leaf—and are set free by the dehiscence of its exterior envelope. They have the property, when brought in contact with the ovule or the stigmatic surface of the ovary, of emitting, through valvular openings in their outer coat, long tubes produced by the outgrowth of the contents and inner wall. The pollen tube passing through the tissue of the style or neck of the ovary to the ovule—or, when this exposed, entering its micropyle directly—comes into that relation with the included germinal corpuscule, which is necesssry to its fertilization.

With these preliminary remarks on the general relations of the reproductive process in the vegetable kingdom, we may now proceed to a summary of the most important modifications of its details in the principal groups, commencing with the simpler cellular forms.

REPRODUCTION IN THE PROTOPHYTA.

Under this head I propose to consider the process in some of the simpler Algæ, which I take by themselves simply for convenience sake, and without hazarding any opinion on the value or limits of the group. The term, I think, is generally employed in this indefinite way, and it is, therefore, to be regretted that its etymology should suggest a parallelism to that of Protozoa, a very natural primary division of the animal kingdom.

I confine myself, then, in this section to Algæ of those

rudimentary forms in which the process of conjugation prevails. All these species may be considered as unicellular; for though the cells are not in every case isolated, they have always the capacity of maintaining an independent life, when by any means they are separated from each other.

When in connection with each other, the cells are arranged in various fashions; in some Palmelleæ they are embedded in a gelatinous matrix, in other cases they cohere sidewise to form flat or bullate fronds, but in the majority they are connected in single file, in filaments simple or branched according to the species. Except in Palmelleæ, the cell wall becomes eventually a well-defined structure, and consists of a layer of cellulose, lined by a film of an albuminous nature, which is probably nothing more than the outer pellicle of the protoplasmic contents. This internal protoplasm is generally coloured green—occasionally red—by an accumulation of granules, termed *endochrome*.

These Algæ multiply in various ways. The simplest is by a process of fission following on gemmation or cell-formation—the points at which the new cells are developed, and their continued adhesion or otherwise, giving rise to the varied forms of differently shaped fronds, or isolated cells. But the most characteristic form of reproduction is by conjugation, already noticed as representing in a rudimentary way that by the co-operation of the sexes.

Conjugation has been observed in the confervoid genus *Zygnema* and its allies, in most of the Desmidieæ, and in some species of Palmelleæ and Diatomaceæ.* The first step in the process appears to be the rupture of the cell wall by the growth of the contents, which then protrude as a secondary or pseudo-cell, surrounded by a pellicle formed within the wall of the original cell. On coming in contact

* For a notice of the recorded observations on the conjugation of the Diatomaceæ, see the Notes to the Translation of Hoffmeister's Papers in the Annals of Nat. Hist., 3d ser. I. p. 7.

with the endochrome, which escapes in a similar way from a neighbouring cell, the two plastic masses unite by the deliquescence at the meeeting point of their imperfect walls and become fused into a single spore, the investment of which subsequently acquires the character of a true cellulose membrane.*

The variations of the process in different cases appear to depend very much on the mode of dehiscence of the original cells. In Desmidieæ, in which the cells—generally isolated—are marked by a median constriction, each of the pairs of cells concurring in the process of reproduction splits into two, and the effused contents comingling in the void space surrounded by the four empty valves, are there developed into a germ mass, whose investing layer acquires subsequently a cellulose character, and usually becomes covered also with siliceous spines.

In Diatomaceæ the pair of conjugating frustules first become embedded in a mass of gelatine, and then the valves of each separate as if hinged at one side, so as to form fissures along their contiguous margins, through which the contents escape, and become amalgamated in the gelatinous matrix. Two sporoid bodies are generally produced, † owing, there is reason to believe, to a sub-division of the original germ-mass. They appear as a pair of new frustules lying crosswise to the original ones, and of much larger dimensions. ‡

In *Meloseira* a somewhat analogous kind of action seems to take place between the ends of a single cell—the cell being an elongated cavity, apparently formed by the fusion of two or three into one. §

In the confervoid filamentous Algæ, Dr. Carpenter gives

* Hoffmeister—Annals of Nat. Hist., 3d ser. I. 1.
† Thwaites in Ann. Nat. Hist., XX. pp. 9 and 343. In *Fragillaria* however, there is but a single spore.
‡ Thwaites in Ann. Nat. Hist. 2d Ser., I. 166.
§ J. H. Carter in Ann., 2d Ser., XVII. 1.

the following account of the process :—" The conjugation ordinarily takes place between the cells of distinct filaments; these approximate to each other, and put forth little protuberances that coalesce, and establish a free passage between the cavities of the cells, whose contents then intermingle."* In reality, however, the outer walls of the conjugating cells are probably as passive in this as in the former cases, the papilla being protruded, and eventually ruptured at the point of contact by the force of growth of the contained endochrome or portions of the wall already beginning to soften. The united contents on their fusion appear to undergo condensation, and the common mass is generally retracted wholly into one of the original cells.†

In the genus *Zygnema*, and in most other species of the group, the conjugation takes place solely or chiefly between the cells of two filaments which lie parallel to each other, so as to present with their connecting passages a ladder-like appearance ; but in some it takes place promiscuously on all sides, interweaving the whole mass of filaments into an inextricable network. There are also species in which the conjugation takes place between the adjacent cells of the same filament : in this case a channel of communication is formed through a protuberance just over the septum, the contents of one cell thus escaping into the other, and then forming the spore ; so that after the process is completed the alternate cells contain spores, the intermediate ones being empty. In all these cases the recipient cell, as the matrix of the spore, may, with a certain propriety, be considered to have a female character ; and when, as in the genus *Spirogyra*, all the cells of one filament thus empty themselves into those of another, the sexual distinction

---

* Principles of Compar. Physiol., p. 190.

† This is the case also in a species of *Didymoprium*, one of the filamentous Desmidieæ. Ralf's Monograph, pp. 58—62.

may, of course, be extended to the whole frond. But in the Desmidieæ and Diatomaceæ, in which the spore is formed outside the cells, and in *Mesocarpus* among the Conjugatæ, in which it is lodged in a dilatation of the connecting passage, we can only in a general way allege indications of sexuality, having no guide to determine which is male and which female.*

The modifications of conjugation just noticed may be arranged as follows :—

I. In isolated cells :
  1. Fusion of two pseudo-cells into one—*Palmoglea*.†
  2. Dehiscence of two cells, and fusion of their contents into a single spore—Desmidieæ.
  3. Effusion of the contents of two cells, and formation from thence of two new frustules, larger than those in conjugation—Diatomaceæ.

II. In cells aggregated into filaments—Conjugatæ.
  4. Contents of two cells in different filaments effused into a dilatation of the connecting tube, and there formed into a spore—*Mesocarpus*.
  5. Contents of the cells of one filament passing promiscuously into those of others, through communicating tubes, the recipient cells maturing the spores—*Mougeottia*.
  6. Conjugation of the same kind between adjoining cells of the same filament—*Rhynchonema*.
  7. Conjugation of the same kind, one filament evacuating the contents of all its cells into those of a connected filament—*Spirogyra*.

---

* Carter remarks that in *Spirogyra* the recipient cell is generally the larger of the two, and hence suggests that the inequality in the size of the conjugating frustules which is occasionally observed in Diatomaceæ may also possibly have a sexual import. Ann. Nat. Hist., 3d Ser., I. 35.

† Braun, Rejuvenescence in Nature, pp. 136 and 327.

8. Conjugation between the contents of opposite ends of the same cell*—*Meloseira*.

The *conjugate-spore* has been seen, in the case of *Spirogyra*, to split its coat in germination in two halves, and emit a fusiform cell, from which others are budded off, so as to form a confervoid filament like that in which the spores were originally formed.† But the development is not always so direct. Recent observations indicate that in various Desmidieæ the contents of the conjugate-spore are transformed by repeated binary sub-division into numerous derivative cells, which finally assume the form of the original cells concerned in the act of conjugation, and are set free by the dissolution of the outer wall of the spore. The existence of such a process was rendered probable by the observations of Focke, Jenner, and Ralfs on *Closterium*, and has since been satisfactorily followed out in *Cosmarium* by Hoffmeister.‡ A similar multiplication of the spore-contents has been noticed in one of the Palmelleæ *(Palmoglea)*, and in *Thwaitesia* and *Mesocarpus* among the Conjugatæ.§

There are other cases again in which the primary product of the spore appears to be a cluster of zoospores, probably reproducing the original form by a second process of germination. This is asserted by Pringsheim of *Chlamydococcus*

---

* This subject is treated at some length by Braun (Rejuvenescence, &c. Ray. Soc. Transl., pp. 284, et seq.)

† This is the account given by Vaucher, Meyen, Smith, and Pringsheim. Agardh, however, maintains that its contents become resolved into Zoospores. The probability of an actual diversity in this respect is confirmed by the observations of Pringsheim and Henfrey, that zoospores are sometimes formed, instead of one large resting spore, as the primary result of conjugation. See Ann. Nat. Hist., 2d Ser. XI., 210-297.

‡ Ralf's British Desmideæ, p. 11 and Pl. XXVII. Annals of Natural Hist., 3d Ser., I. 16.

§ Micrographic Dictionary. See also Berkeley's Introduction to Cryptogamic Botany, p. 152, and Prof. Braun's Rejuvenescence in Nature. Ray. Soc. Transl., p. 136.

and some Palmelleæ.* In *Palmella* itself Mr Thwaites figures a branched filament proceeding from a spinulose cell—very like a conjugate spore—and bearing ordinary palmella-cells at its free extremities.† And so in Diatomaceæ, from the frustules developed in conjugation being much larger than their parents, it is probable that they also split up into smaller pieces or frustules.‡

The conjugate spore, whatever be its mode of development, is always very retentive of vitality, and is commonly several months of germinating. Hence it is fitted to perpetuate the race from year to year, and is frequently spoken of as the *resting* or *winter-spore*. Bodies of a similar kind are found even in tribes in which as yet no conjugation has been observed, but in these cases we have no satisfactory information concerning their mode of origin. The dissemination of the species under circumstances which admit of continuous vegetation, is effected by *zoospores* or motile gemmæ, very similar of those of the higher Algæ.

Remarkable as the process of conjugation undoubtedly is, and quite different from all other modifications of the reproductive process in the vegetable kingdom, it is very doubtful if the species in which it occurs can be definitively marked off from others, as constituting a natural group of themselves, for in many, which certainly seem to be closely allied to those now under consideration, fertilization appears to be effected, as in the higher algæ, by *phytozoa*. Corpuscules of this nature are described by Carter in the remarkable unicellular alga *Volvox*, which ranked so long as an infusory animalcule.§ In other cases again, as in *Cylindrospermum*, among Nostochineæ, the filaments are termi-

---

* Quar. Journ. Micros. Science, IV., 313.
† Ann. Nat. Hist., 2d Ser., II. 313.
‡ Thwaites' Ann. Nat. Hist., XX. (1847). Griffith Ann. 3d Ser. XVI. 92, 1855.
§ Annals of Nat. Hist., 3d Ser., II., 237, III. 1.

nated by a pair of cells, of which the proximal forms a spore in its interior, while the distal, which goes under the name of *heterocyst*, has been considered to contain spermatic matter, though no motile particles have been seen to issue from it, nor anything of the nature of fusion or conjugation been observed to take place between it and the spore-cell. Clear cells, to which the name of heterocysts has also been applied, occur in most other Nostochineæ, but without the accompanying sporangia, so far as has yet been observed.* There are besides other species in which neither conjugation, nor indeed any form of digenesis, has yet been detected. This is the case, for instance, with the Oscillatorieæ, and Ulvaceæ, which are not known to multiply but by the detachment of cells or frustules, and in the case of the latter also by the emission of zoospores. It is asserted, however, by Itzigsohn that *Oscillatoria* is only a particular phase of *Chlamydococcus*, a unicellular alga allied to *Palmella*—the filaments of the former arising from minute spiral corpuscules, generated within green gonidia discharged from the latter; while the frustules of the *Oscillatoria* are stated to be again developed into the globules of *chlamydococcus*.† The subject at least invites farther investigation, for there is a prevailing impression among botanists that the Oscillatorieæ are merely barren phases of plants which may fructify in some other form. Among the Ulvaceæ it is to be noticed that in some species motile corpuscules occur of two kinds, borne on different plants, one *quadriciliate*, the other *biciliate*. The former are regarded by Robin as phytozoa, but Thuret alleges that both kinds germinate. If phytozoa exist we should expect to meet also with corresponding bodies of the nature of resting-spores, but such have not yet been detected, at least in the majority of the group.

---

* M. G. Thuret—Ann. Nat. Hist., 3d Ser., II. 1.
† Quarterly Journal of Micros. Science, II., 188 (Botanische Zeitung).

There is reason to believe that future researches will show that conjugation, or some cognate process, is of very general occurrence among these low forms of organization, for if any modification of digenesis may be held to extend to the most rudimentary forms of living matter, it is most likely to be that in which there is little or no differentiation of sex. Alternation of form, too, must prevail to some extent, if the suspicions of naturalists be correct as to many of the reputed species being merely transitory phases of others. At present, however, all is conjecture on the subject, and the only well-ascertained phenomena of the kind are those just noticed as occurring as a preliminary (or protomorphic) stage in the germination of certain spores.

A tabular view is here subjoined of some points in the development and reproduction of the orders of algæ just referred to :—

IN THE VEGETABLE KINGDOM.

TRIBES.	VEGETATION.	MONOGENESIS.	DIGENESIS.
Palmellæ	Unicellular	Zoospores (in some*)	Fusion of two cells.
Desmidieæ	Unicellular or filamentous	Swarming Granules†	Conjugation with single spore.
Diatomaceæ	Unicellular (in general)	Zoospores doubtful‡	Conjugation with double spore.
Conjugatæ	Filamentous	Zoospores general	Conjugation with single spore.
Oscillatorieæ	Filamentous	Z. in palmellean phase ?	Conjugation in the same?
Ulvaceæ	Frondose	Gonidia of two kinds,	One possibly spermatic.
Nostochineæ	Gelatinous	Endogenous cell formation	Sporangia and heterocysts.
Volvocineæ	Unicellular	Formation of inner spheres	Sporangia and phytozoa.

* Braun and Cohn in Bot. and Physiol. Mem. of Ray Soc.   † Ralfs' Desmidieæ, p. 9.   ‡ Braun Op. Cit. p. 139.

4. § REPRODUCTION IN THE ALGÆ GENERALLY AND CHARACEÆ.

Of the three current divisions of Algæ—Confervoideæ or green-spored, Florideæ or red-spored, and Fucoideæ or olive-spored—the first is generally considered as including those simpler forms already noticed in which conjugation occurs, as well as others of higher organization, in which, as in the order generally, a distinction of sexual elements seems to prevail.

The spermatic particles of Algæ go under the name of *antherozoids* or *phytozoa*, but, with one or two doubtful exceptions, they are not filiform like the bodies so called in the higher Cryptogamia, but ordinarily of an ovoid form, with two long cilia attached at one end, whose play impresses upon the corpuscule a rapid jerking motion. The particles supposed to play the part of antheriozoids in the Florideæ appear to be destitute of cilia and of all motile power. The germinal bodies are small globular masses of protoplasm occupying the interior of cells termed *sporangia*. No conjugation or direct union of cells takes place in this group, but the concourse of the elements is effected by the formation of pores in the cell-walls, through which the spermatic particles escape from their proper cells and gain access to the interior of the sporangia; the perforations in the latter are termed *micropyles*.

Zoospores—a sort of motile ciliated gemmules found in connection with the conjugating Algæ which have no antherozoids—occur also throughout the present group, except among the Florideæ. Though liable to be confounded together, the two kinds of corpuscules present generally certain structural diversities, and have totally different functions. The zoospores are generally larger, their cilia are more numerous, and their motion through the water more uniform. They appear to be formed by the breaking up of the endochrome of some of the component cells of the filament or frond, and escape by perforations in the walls; the antherozoids again originate in special cells of the nature of *antheridia*. They are both indeed concerned in propagation, but while the peculiar office of the latter is to impregnate the germs, the zoospores appear to be merely a kind of gemmæ capable of spontaneous development, for it is observed that after a time their motion ceases from the loss of the cilia, and they begin to germinate into new fronds.

Only such particulars of the reproductive process in the minor divisions of the order will be noticed here as are illustrative of the main points now under consideration. To begin with the Confervoideæ—as limited to the green-coloured Algæ, both fresh water and marine, which do not conjugate—we may take for illustration of the filamentous species, formed by the coherence of cells in linear series, those contained in the genera *Sphæroplea*, *Bulbochæte* and *Œdogonium*, as having been the subjects of the most satisfactory observations. Zoospores have been detected only in some of the species, but the occurrence of antherozoids appears to be general. They are developed from special cells, either directly as in *Sphæroplea*, or through the medium of a prothallial frond or androspore, as in various species of the other two genera. In the latter, according to the observations of Pringsheim, small bodies like zoospores *(microgonidia)* are formed singly in cells, which are smaller than the rest, and present certain other peculiarities. These corpuscules, after their period of activity is over, attach themselves to the neighbourhood of the sporangia, and germinate into minute fronds of three cells, one of which serves as a pedicle of attachment, while the other two become antheridia, each maturing an antherozoid of some size. By a peculiar fissuring of the antheridium, and the formation of a pore or micropyle in the sporangium, the antherozoid passes from its own cell to that containing the globular mass of germinal matter, in the substance of which it seems to be absorbed. The mass then acquires a membrane of cellulose on its exterior, and becomes a true spore.*

In *Œdogonium*, *Bulbochæte*, and *Coleochæte* a process has also been observed during the germination of the spore, which might be termed one of protomorphic gemmation.† In these, and perhaps also in some other genera, the spore does not germinate at once into a new frond, as in other Algæ *(Achlya, Vaucheria, &c.)* but becomes converted into a sort of capsule containing four bodies, precisely resembling the ordinary zoospores of the plant in their appearance, their motion, and their power of germinating into new fronds.

The genus *Vaucheria* may be taken as the type of another family of confervoid Algæ—the Siphoneæ—which are sometimes described as unicellular, the filaments having no tranverse septa. The sporangia commence as lateral diverticula, but are afterwards shut off as separate cells. The antheridium originates as a papilla, in close proximity to the sporangial dilatation, and generally precedes it a little in its first development. It ultimately acquires the form of a hooked process, or abortive tendril ("hornlet"), curving over the sporangium. The point of the hornlet is soon converted into a separate cell, and its contents, which had previously become colourless, are then resolved into antherozoids. Perforations taking place subsequently, both at the point of the hornlet, and at the most prominent part of the sporangium, the antherozoids come in contact with

---

* Pringsheim's Report to the Berlin Academy. See Quarterly Journal of Microsc. Science, IV., 131, and Edinb. Philosophical Journal, N.S. V., 376 (1857). In the species of *Œdogonium* observed by Mr Carter the antherozoids were not formed in free androspores, but were emitted directly by peculiar short cells, grouped in pairs or clusters of three, and giving a ringed appearance to certain parts of the filament. Ann. Nat. Hist., 2d Ser., XVIII., 81, and 3d Ser., I., 19.

† Edinburgh Philosoph. Journal, V., 376 (1857). Quarterly Journal of Microscopic Science, IV., 133.

the unwalled contents of the latter, which are then converted into a spore, the soft mucous investment being transformed into a consistent cell-wall.

The latest observations on *Vaucheria* we owe to Pringsheim,* who has also directed much attention to the corresponding processes in *Achlya (Saprolegnia* Kutz.*),* a genus really closely allied, though of very different external appearance, the filaments being colourless and much more minute, and growing like a tuft of mouldiness on dead flies and other animal matter macerating in water. The sporangium has here the same general relations as in *Vaucheria*, but their walls are perforated by a great number of pores, and they contain eventually numerous spores. Although no antherozoids have yet been detected, the antheridia appear to be represented by slender vermiform branchlets which arise between the sporangia, and sometimes twine round them in irregular coils. These, according to Pringsheim, throw out lateral papillae, which protrude into the pores of the sporangia.†

Both these genera propagate usually by gemmae developed in the dilated extremities of the branches, which are converted into cells by the formation of transverse septa in the filaments. In *Vaucheria* the whole contents, when mature, are thrown out through a perforation in the apex of the cell, as one large ciliated spore; in *Achlya* they break up into a swarm of motile corpuscules like the zoospores of other Algae. It is not ascertained whether the co-existence of these two kinds of reproductive bodies is connected with any form of alternation of generation; no phenomena of the kind have yet been observed in this group.

* Quart. Journ. Micr. Sc., IV., 63, 130.
† Braun, Nægeli, and Karsten supposed that a real conjugation of cells took place in Siphoneae; but the view founded on the late observations of Pringsheim is now generally received. *Achlya*, however, is still considered by some as merely the early phase of a true mould or fungus.

Among the Florideæ or rose-coloured algæ—a group exclusively marine—we find commonly three different kinds of fructification, situated mostly on separate fronds :—

1. *Tetraspores*, considered as gemmæ by Thwaites, Harvey, Berkeley, and Pringsheim, and consisting of quaternary aggregations contained in receptacles, which are in some cases simply imbedded in the frond.

2. *Antheridia*, containing, in special receptacles, particles which have been taken for antherozoids, though destitute of cilia, and perhaps also of motile, power.

3. *Conceptacles*, containing clusters of sporoid bodies, which are generally regarded as the germs or true spores.

Pringsheim, however, is inclined to think that the last-mentioned bodies—at least in some species of *Ceramium*—are rather gemmæ, which develope prothallia for the production of the real germs, as he has seen them begin to germinate while still in the conceptacle, before any perforation had taken place for the entrance of antherozoids, and observed that the result was an irregular cellular growth, quite different from the ordinary frond. In this view the prothallium, though unisexual like that of *Œdogonium*, would differ in representing the opposite sex, bearing not spermatic but germinal particles, whose impregnation would be effected by the rod-shaped antherozoids formed directly on the male frond.

The constancy of the occurrence of three forms of reproductive organs, just noticed on different plants, gives a certain probability to the idea of an alternation of forms, like that occurring in some species of animals—the fronds originating from fertilized germs bearing only gemmæ (tetraspores), which, in turn, would develope other fronds maturing true sexual organs.

In most of the littoral group of fucoids or olive-coloured seaweeds, antherozoids and proper germs have now been detected, generally in separate receptacles on the same frond. According to the observations of Thuret on different species of *Fucus*, impregnation is effected by the simultaneous evacuation of both elements from their respective envelopes. During the recess of the tide, when the tissues contract somewhat by desiccation in the air, the antheridial cells are expelled, as well as the contents of the so-called spores, consisting of eight protoplasmic bodies enveloped in a gelatinous mucus. On being again submerged, the antheridia rupture by imbibition, and the antherozoids come in contact with the germinal bodies. In about twenty-four hours after this, the gelatinous layer, which surrounds each germ mass, is converted into a tough membrane, in which Pringsheim mentions observing certain red corpuscules, which he considers to be the nuclei of the antherozoids.* Then follows germination—a simple process of cell-formation, resulting in the production of a new frond. In some fucoids, however, the cells first formed in this process separate from one another, being set free by the rupture of the common investing membrane of the spore ; and each of them may give origin to a distinct plant or phytoid.

Zoospores are now known to occur in this group also, but no obvious alternation of forms has ever been observed.

It would seem, therefore, that the only general conclusion in regard to the reproduction of the Algæ which, in our present defective state of information, we are warranted in drawing is this : that a sexual act probably occurs normally from time to time in the life-history of all the species, though impregnation has been actually observed only in a few. We know, indeed, that non-sexual propagation prevails to a great extent and in great variety of forms. Not only is there the usual pullulation of new shoots, characteristic of vegetables generally, which is here represented by the

* Journ. of Micros. Science, IV., 125 (An. des Sciences Nat.)

continued growth and occasional branching out of the frond by the addition of new component cells, but independent fronds are also sometimes produced by detachment of bulbils or other portions of the substance of a pre-existing one ; and, besides this, almost all the species bear more than one kind of sporoid bodies. We are not, however, in a position to say whether, or in what way, spore-gemmation alternates with sexual generation generally throughout the group. We only know that in a few species a *gamomorphic* porthallium is budded off from the spore for the development of the reproductive organs of one of the sexes, and that some of the same species *(Œdogonium* and its allies) are also the subjects of a sort of gemmation in the protomorphic stage, as well as certain fucoids—the spores which actually germinate into new fronds being only a secondary product of impregnation ; while in other cases (some Florideæ as *Ceramium*) there are certain theoretical grounds for supposing that the sexual fronds may originate from the tetraspores of the gemmiparous fronds.

A few words may be added in reference to the order Characeæ, on the systematic position of which botanists are not quite agreed. These plants are described by Berkeley as consisting " of confervoid articulated threads, simple as in *Cladophora*, or compound as in *Polysiphonia*."[*] In their vegetative axis, therefore, they have quite the organization of the true Algæ, and though their reproductive organs are certainly very peculiar, they are perhaps reducible to a modified form of the same general type. Two kinds of fructification are met with in *Chara*, the *nucule* and the *globule*, which are supposed to contain respectively the germinal and spermatic elements. This opinion rests mainly on the development in the latter of articulated filaments, containing motile ciliated bodies, having a great general resemblance

[*] Cryptogamic Botany, 425.

to the antherozoids of the higher Cryptogamia, for impregnation has never been actually observed. The globules of *Chara* are round red bodies, whose outer wall consists of eight pieces, each having the form of an equilateral right angled spherical triangle, and consisting of a whorl of cells, radiating from the outer extremity of a perpendicular column. The supporting columns of all these parietal segments arise in the centre of the globule, from an inward prolongation of the footstalk of the organ, and at the same point there are also attached numerous jointed threads, each of the articulations of which produces a spiral antherozoid, with two cilia at its dilated extremity. Mr. Berkeley ingeniously explains the morphosis of the globule as a fascicle of branchlets given off from the tip of the axis, radiating in eight different directions, and each producing in turn another whorl of branchlets, which, by the co-aptation of their extremities, form the wall of the globule.*

In connection with this the antheridial threads may be considered as a secondary order of filaments, arising in the axils of the first, and bearing, in their individual cells, the spermatic particles—a view which helps to approximate the fructification to the type of antheridial development among the Algæ.

The co-related organ—the nucule—consists of a central sac filled with starch, and coated by a layer of five elongated cells, wound spirally round it, attached at their bases to the insertion of the footstalk, and free at their tips, where there is an aperture leading into the interior. Morphologically, therefore, the nucule may be looked upon as a whorl of unicellular branchlets, cohering by their edges, but without the secondary fascicles, and the axillary antheridial filaments, which give such a complex character to the globule.

The nucules germinate by the formation of a cell at the

* Cryptogamic Botany, 127.

top of the central sac, which developes downwards into rootlets and upwards into stem.*

The Characeæ multiply also by deciduous leaf-buds ; but we are very deficient in observations regarding the development of this singular group of plants.

5. § REPRODUCTION IN FUNGI AND LICHENS.

The Fungi are a large order of aerial cellular plants of low development, intermediate between Algæ and Lichens, and passing into both orders by the closest possible affinities. In their structure there is generally a very marked distinction between the vegetative tissue and the reproductive organs ; the former, termed the mycelium, is always the most inconspicuous, and frequently escapes observation altogether, being embedded in the soil or basis on which the fungus grows. It consists of a mass of branched confervoid filaments, inextricably entangled, and frequently even anastomosing together.

The vitality and vegetative power of these filaments is so great that even fragments of them will suffice to reproduce the mycelium. Such fragmentary portions are indeed profusely detached as gemmæ in the normal development of the tissue ; and as the particular form and connections of these reproductive bodies vary at different periods of the life-history of the plant, we meet with a great variety of fructification even on the same species ; as many as five different forms are stated, on the authority of Mr Berkeley, to occur in the same species of *Erysiphe*.† As the specific identity of these forms, depending on their connection with the same mycelium, may readily escape observation, it is not wonderful that many of them figure as distinct plants

* Berkeley, Op. Cit., p. 428. Carter in Ann. Nat. Hist., 2d Ser., XVIII., 107.
† Cryptogamic Botany, p. 78. Quarterly Journal of Microscophic Science, Jan., 1857, p. 51.

in scientific works. The consequence has been the establishment of a great number of spurious genera and species, and the inextricable confusion of a subject, involved from its very nature in much perplexity. Thus the numerous species assigned to *Sclerotium* are in reality perhaps none of them proper or *autonomous*, but only transitory conditions of Fungi already known by other names, such as *Peziza*, *Sphæria*, *Agaricus*, &c., which in maturity differ as widely as can well be conceived.*

Great progress, however, has lately been made by the researches of MM. Tulasne and others in tracing out the reproductive process in this group ; and, though we are still far from being in a position to draw positive conclusions, there appear to be good grounds for the opinion that organs of sexual import are very generally present. There is, indeed, a general agreement as to the male element being represented by the *spermatia*, minute staff-like bodies, somewhat resembling the supposed antherozoids of the red-spored algæ. These particles have no cilia nor any motile powers beyond what may be accounted for on physical principles, and the received views as to their functional import rest less on any positive evidence than on analogy, and on their not possessing any power of germination. They have not been met with, save exceptionally, in the higher Fungi, but have now been very generally detected in other divisions of the order. They are usually attached to filaments which are either associated with the organs bearing the fertile spores, or occupy distinct receptacles.†

The true spores, which have been seen to germinate, are developed in the interior of cells, or are attached to their exterior. External or naked spores are met with both in the Hyphomycetes or moulds, and in the higher divisions of

---

* Berkeley's Cryptogamic Botany, p. 268.
† Tulasne, Comptes Rendus. March 31, 1851. See also Annals of Nat. Hist., 2d Ser., VIII., 117.

Gasteromycetes and Hymenomycetes. In the first they occur in beaded chains clustered at the ends of filaments; in the latter they are attached in groups of four to the apices of elongated cells, termed *basidia*. When the spores are developed in the interior of cells, the latter— which are termed *asci* or *thecæ*—have generally the form of elongated pods. They come to maturity at a later period than the spermatia, and then contain a variable number of spores—usually some power of two—arranged in linear series. The presence of spores of this kind *(ascospores,* or *thecaspores)* has long been recognized in one large tribe of Fungi, hence called Ascomycetes ; but the researches of Tulasne go to shew that they occur also in other divisions of the order. In the large group of Coniomycetes, including the smuts, and other such like epiphytic parasites, it has been shown that many of those Fungi which produce naked spores are but preliminary stages of species, which, in another phase of development, form spermatia, and spores contained in asci, or at least reducible to that type. Thus the epiphytes, which go under the name of *Uredo*, are the protomorphic or rudimentary forms of others, hitherto referred to distinct genera, as *Æcidium, Phragmidium, Puccinia*, &c. *Stilbospora* in the same way appears to be a precursor of a form of *Sphæria*.* Mr. Berkeley observes that "it is quite certain that a large portion of the so-called species of *Phoma, Leptostroma, Diplodia, Hendersonia, Cytispora, Septoria*, &c. are mere cases of dualism ; and the same may, without much chance of error, be predicated of those cases

* Tulasne, Ann. Nat. Hist., 2d Ser., VIII., 114. Comptes Rend., Mar., 1851. Currey in Philos. Transac. for 1857, p. 548. M. L. R. Tulasne, along with many others, identifies also the following :— *Næmaspora Ribis* with *Sphæria Ehrenbergii, Micropera Drupacearum* with *S. Leoeillei, Asteroma Ulmi* with *Dodithea Ulmi;* and he remarks that in *Rhytisma* every species may be said to have a precursor in a *Melasmia* (a fungus with acrogenous spores), which play the same part here that *Cytispora* and its analogues do in regard to the *Sphærias*.

as *Dilophosporium*, *Neothosporium*, and *Pestalozzia*, where the objects are of some interest on account of their curious appendages. All, indeed, are interesting so far as ascertained dualism is concerned, or as far as there may be a prospect of shewing that they are the spermagonia or pycnidia of ascophorous species."*

This relation of associated forms may be illustrated by the development of *Æcidium*. The spores of this fungus produce a mycelium, the filaments of which bear the naked spores of the *Uredo*, and these, when they come to germinate, originate a second mycelium, in connection with which are developed spermatia and the thecaspores of the *Æcidium*.†

It is possible that in some species without spermatic fruit particles of a similar kind may be developed from certain of the spores, by an arrangement like that described by Pringsheim in the androspores of *Œdogonium*, for in the germination of certain fungi it has been observed that while some of the spore-like bodies emit mycelial threads, others give origin to minute corpuscules, which undergo no farther development. These have been conjectured to be of a spermatic nature.‡ In some species of *Peziza*, according to Radlkofer, a process of this kind occurs, while in others the corresponding bodies originate the spermatic particles less directly by the intermediate development of a special spermatiferous mycelium.§

This may be considered as a sort of alternation of generations; and it is to be observed that it occurs in the gamomorphic stage of development, that is, in the maturation of

---

* Introd. to Cryptogamic Botany, p. 331.

† Tulasne, as quoted before. Sometimes the succession is complicated by other intermediate forms; thus in the germination of some species of *Erysiphe* a succession of mycelial forms are produced from sporoid bodies of different kinds, and the series is closed by the development of the sporiferous thecæ.

‡ Currey in Journal of Micros. Science (April, 1857), pp. 124-126.

§ Radlkofer—Ann. Nat. Hist., 2d Ser., XX., p. 247.

the sexual elements, in which respect it contrasts with a process of multiplication occurring in some other cases in the protomorphic stage—viz., the binary division and sub-division of the contents of the spore into so many reproductive cells, that the original spore becomes itself a sort of secondary theca. All these derivative cells may germinate, and in some cases they do so simultaneously, sending out their mycelial filaments through the outer or common wall, so as to give the composite spore somewhat the appearance of an insect with its legs extended on each side of its body.* As might be expected, an isolated fragment of such a compound spore still retains the power of germinating.

In illustration of these statements the development of the ergot-fungus may be referred to, as explained by Tulasne and others, though some points still remain to be cleared up, for the completion of its history, and the harmonizing of the accounts of different observers. The fungus originates as a mycelial growth in the ovary of the rye or other grass which it attacks, and soon matures its *conidia* or naked spores, on the ends of filaments. These have been observed by Tulasne to germinate and become elongated into new filaments, which appear to be concerned in the formation of the dense fibro-cellular stroma of the proper ergot-growth *(sclerotium)*. This gradually acquires the form and dimensions which have led to the grain affected receiving the name of "spurred rye." The ergot-growth, when shed and exposed to moisture, gives origin to a small club-shaped fungus, to which the name of *Claviceps*† has been given, and which is identical with a species known to botanists as *Sphæria purpurea*. This developes on the surface of its globular heads numerous small cavities or conceptacles, with thecæ in their interior, which bear spores

---

* Currey in Quar. Jour. Micr. Sc., IV., 200; also April, 1857, p. 122.
† *Cordyceps* or *Cordyliceps* of Fries.

in the usual way. The sporoid bodies, when they gain access to the flower of the rye, develope the fungus-growth, from which the ergot originates in the way described.*

The point of greatest difficulty is the determination of spermatic organs. Spermatia do not seem as yet to have been detected in connection with the sphæria-growth. Léveillé and others appear to regard the conidia as "fertilizing organs;" but besides that this is not consistent with Tulasne's observations of their germination, the wide interval between them and the spores is of itself opposed to such a view.

The highest group of Fungi is the one in which our knowledge of the reproductive process is as yet the least satisfactory. Spermatia have been met with only exceptionally—as in Tremellineæ†—and proper thecæ, as it would seem, not at all. The characteristic fructification consists of naked spores borne in clusters of four, as described already, on elongated cells termed basidia, which in the Hymenomycetes occupy some part of the exterior surface of the proliferous receptacle—as we see in the gills of the common mushroom—while in the Gasteromycetes they are developed from the lining membrane of an internal cavity of which we have an example in *Bovista* or the puffball. In this case the spores

---

* Comptes Rendus. Dec., 1851, and Seq. Ann. des Sciences Nat., 3d Ser. XX., p. 553. Annals of Nat. History, 2d Ser. IX. 494. Compare the account in the Micrographical Dictionary.

† In some of this group, according to Radlkofer, the *hymenium* or fructifying surface bears spermatia, and basidia crowned with spores like those of agarics. This occurs in *Tremella mesenterica*. In others, as *Dacrymyces deliquescens*, we have in the same position multilocular spores, all of like aspect, but some acting as androspores, and originating in germination spermatiferous pedicels; while others, representing apparently the opposite sex, emit mycelial filaments. Annals of Natural History, 2d Ser., XX., 247. See also Berkeley's Introduction to Cryptogamic Botany, p. 350.

when ripe escape by special apertures, or by the general breaking up of the cyst. The basidia with their quaternate spores have certainly much more the character of a gemmiparous fructification, such as the stylospores of the lower Fungi or the tetraspores of the Florideæ, than of impregnated germs, such as the spores produced in thecæ are presumed to be ; so that in these plants—though they seem to stand at the head of their own type of vegetation—the only obvious reproduction is of a non-sexual kind. It remains for future investigation to decide whether sexual organs exist at all, or what may be their relations ; whether the spores, like those of the ferns, originate androgynous prothallia, or whether the stool of the mushroom, like the capsule of the moss, is itself the result of a process of impregnation, effected by organs standing in some more immediate relation to the mycelium, which have not yet been detected. That an " alternation" of some kind prevails the analogy of other Fungi would lead us to anticipate.

In the group of Hyphomycetes also, which presents one of the lowest types of fungoid growth, containing the various forms of mould and mildew, spermatia, and spores fecundated by them have not yet been generally recognized. But there are grounds for believing that many of these vegetations are not true or autonomous species, but, like the forms of Coniomycetes before noticed, merely states of the mycelium of other species, bearing a secondary kind of fructification. Thus *Aspergillus* has been observed by Bary to yield a form of *Eurotium*; *Fusarium* is known to develope a species of *Peziza*; the genera *Tubercularia* and *Myriocephalum* appear to be mere precursors of species of *Sphæria*; and *Oidium* is admitted generally to be only an early condition of *Erysiphe*.* Others of these rudimentary forms of organization, it is possible, ought rather to be referred—

---

* Berkeley Op. Cit., pp. 218, 292, 300, 330.

along with the lower algoid forms, with which they have close affinities—to the group of Protophyta. When the mycelium originates, as is frequently the case, in a watery fluid, it so precisely resembles the filamentous growth of some of the algoid Protophyta, that we find many of them described as such in botanical works. Thus the white-felted mass, which is apt to form in ink, is described as an Alga under the name of *Hygrocrocis atramenti*, but if its progress be watched, it is clearly seen to be only the submerged mycelium of some common form of mould, and it is the same with the so-called "vinegar plant." The characteristic fructification appears only on those branches which emerge from the fluid, and generally consists of naked sporules attached in various fashions to the ends of elongated filaments. Even the genus *Achlya*, commonly regarded as a well-established alga, and having certainly many points of analogy with *Vaucheria*, as already noticed,* is regarded by Berkeley as only a fluid-born mycelium of *Mucor*, a well-known variety of mould.† This view is hardly reconcilable with the presence of sexual organs in *Achlya*, which the researches of Pringsheim seems to indicate. It is to be remarked, however, that these observations do not actually demonstrate the existence of spermatic particles in the organs, which, from their analogy with those of *Vaucheria*, he was inclined to consider as antheridia.

If any of the group of Fungi now referred to are rightly to be considered as allied to the Protophyta, we should look for some modification of conjugation, either in the submerged or the aerial filaments ; and it is so far in support of such a

---

* V. Infra., p. 108.

† Berkeley's Cryptogamic Botany, pp. 132, 148, 295. The authors of the Micrographic Dictionary mention (Art. Mucor) that their experiments have hitherto afforded only negative results.

view, that a process of this kind is known to occur in the genus *Syzygites*, a mould common on decaying agarics.*

The order of the Lichens may be noticed here on account of their close analogy to the ascophorous Fungi in their organs of fructification. The reproductive organs, known under the name of *apothecia*, have long been recognised; they have generally the appearance of shield-like discs of a different colour from the rest of the frond or thallus, and contain sporiferous thecæ, intermingled with filiform or clavate processes (abortive thecæ), termed *poraphyses*, both placed in a perpendicular position, and forming the coloured layer of the apothecium. The co-related organs are of later discovery, and are termed *spermagonia*. They consist of small cavities in the thallus, indicated externally by punctiform apertures, serving for the emission of the spermatia, which exactly resemble those of Fungi, and are produced like them on the ends of filaments projecting from the walls of the cavities.†

Besides these two forms of fructification many Lichens possess other minute organs, called *pycnidia*, outwardly resembling spermagonia, but containing corpuscules termed *stylospores*, which differ from spermatia in being usually oval or pyriform, hollow, and of large size—that is, altogether more like spores. In some species (as *Scutula* and *Abrothallus)* these pycnidia seem to replace the spermagonia, on which account, and from the existence of certain intermediate forms, Dr Lindsay is inclined to consider them as modifications of these organs.‡ But as organs known as pycnidia co-exist in certain Lichens with spermagonia of the ordinary character, and in some cases yield corpuscules

---

* Griffith & Henfrey's Micrograph. Dict., p. 626.   Berkeley's Cryptogamic Botany, p. 295.

† Tulasne Comptes Rendus.   March 24, 1851.   Ann. Nat. Hist. N.S., VIII., 114, XXXII., 427.

‡ Quarterly Journ. of Micros. Science (Jany., 1857) p. 53.

capable of germinating, the stylospores, in these cases at least, appear rather to be—like the bodies of the same name among fungi—a secondary form of spores.* They seem to occur with greatest frequency in species most allied to the Fungi.

In many Lichens we find yet another form of organs, which are undoubtedly of a gemmiparous kind; these are the *soredia*, little pulverulent masses of the green cells termed *gonidia*, of which, generally throughout the order, a stratum is interposed between the medullary and the upper cortical layer of the thallus. The beaded filaments formed by these gonidia are probably the only organic peculiarity which separates the order of Lichens from certain tribes of Fungi, for between these two orders we find a great general similarity, both in the parts of fructification, as now noticed, and in the structure and disposition of the tissues. In both we have a primordial structure or substratum of confervoid filaments, known as the *mycelium* of the fungus, and the *hypothallus* of the lichen; in both also we have generally, but by no means universally, a more compact tissue—the *hymenium* or *crust*—developed subsequently to the other, and in more immediate connection with the organs of reproduction.

No phenomena of the nature of alternation are as yet known to occur among Lichens.

§ 6. REPRODUCTION IN HEPATICÆ AND MOSSES.

These orders introduce us to that higher division of the Cryptogamia, which is characterized by the formation of a leafy axis. Such an axis first appears in the section of the Hepaticæ represented by *Jungermannia*; it is universal in mosses, and attains its maximum development in ferns, where it is constituted in part of fibrovascular tissue, while it is entirely cellular in the lower forms.

---

* Dr. Lindsay in Edin. Philos. Journ. (July, 1859), p. 124.

As reproductive organs, both Hepaticæ and mosses bear, on the same or on different plants, certain correlated structures, termed *antheridia* and *pistillidia*, and performing the same sexual functions as the anthers and pistils of the higher plants. They are frequently surrounded by a leafy sheath, or a circle of leaf-scales, termed the *perichætium*, and are either immersed in the cellular tissue of the receptacle, or attached to its surface, or raised on a peduncle.

The antheridium is a small club-shaped process, which contains in its distended extremity minute cellules, each enclosing an actively moving filament or antherozoid. The pistillidium, or *archegonium* again, consists of cells built up into a flask-shaped structure somewhat resembling the pistil of a flower, with a dilatation like an ovary at its base, containing a peculiar central corpuscule. Through the canal in the apparent style some of the antherozoids seem to gain access to the corpuscule, and the latter on fecundation developes, by endogenous formation, a whole mass of cells which eventually assume, by a process of transformation, in the course of growth, the form of the theca or capsule. In mosses this organ, with its seta, calyptra, operculum, and columella, has rather a complicated structure; in Hepaticæ it is simpler, but in all cases alike its internal cellular tissue developes a mass of dust-like spores. In the frondose Hepaticæ the spores germinate at once into the lobed expansion of green parenchyma, constituting the typical form of these plants, but in those species which have a true axis, and in mosses, the germination first gives origin to confervoid threads, which, after ramifying into a mass of tangled filaments, termed the *protonema*,* send up here and there a leafy axis, bearing eventually like the original one antheridia and archegonia.

This development of the fecundated germ into the organ

---

* Once ranked as an alga under the name of *Byssus*.

known as the "fruit"—which, whether more or less complicated, has always this remarkable peculiarity, that it is an apparatus for the multiplication of germinating particles or spores—has been compared by some authors with alternation of generation. The analogy may at first appear far fetched, as there is no actual separation of parts, the new structure seeming as much an appendage of the original leafy axis, as the fruit is of any phænogamous plant, yet closer investigation will certainly show a real community of nature between the two processes. As this point, however, will again come under consideration, no farther reference need be made to it at present than to remark that in this case two of the stages of development, which were distinguished in the last chapter, are very distinctly marked :—

1. That termed protomorphic, immediately following the fertilization of the germ, by which are formed the "fruit" and the derivative spores.

2. That termed orthomorphic, which consists in the development from the latter of the typical plant, bearing in turn new reproductive organs.

The point mainly to be observed is this, that the fecundated germ is not at once developed into the typical plant, but into an intermediate form known as "the fruit," in which numerous spores are formed. These spores, in germination, give rise to the typical form, either directly or through the intervention of a filamentous protonema, and the typical frond or axis bears the reproductive organs, with the sexes, sometimes united, sometimes separated. In general the orthomorphic form is much more conspicuous than the protomorphic. *Buxbaumia* is perhaps an exception, as the axis which supports the archegonium and afterwards the capsule is here represented only by a tuft of minute scales. The protonema is perhaps best regarded as a second link in the series of forms marking the protomorphic stage.

For the normal fructification of mosses a certain dryness and insolation are required. In a moist and gloomy atmosphere the stems, instead of bearing fruit, are converted into leafy shoots, from which other like shoots may be budded off, the pullulation being continued indefinitely, and the development of fructification postponed, till some change occurs in external circumstances. For a time the pullulations continue in adhesion to the parent stock, forming a ramose vegetation, but the common axis being of a perishable nature they are eventually set free by its decay. In this way even mosses which rarely fructify may extend themselves very widely, rivalling, as has been remarked, the largest forest trees, both in the space covered by their derivative phytoids, which physiologically are parts of themselves, and in their aggregate longevity. Free gemmæ, also, which are detached at once to form new plants, are not uncommon, both among mosses and Hepaticæ.

§ 7. REPRODUCTION IN FILICES AND EQUISETACEÆ.

In these orders we have antheridia and archegonia, bearing a general resemblance to those of mosses; they are not formed, however, in connection with the leafy axis, which bears only the gemmæ known as spores, but in minute detached parenchymatous phytoids, termed *prothallia*, into which the spores germinate, and from which the embryo emerges as the result of impregnation.

Two views may be taken of the homologies subsisting between these plants and the mosses. A physiological correspondence being admitted by all between the gemmiparous production of spores in both cases on the one hand, and the formation of antheridia and archegonia on the other, some eminent cryptogamists — as Hofmeister and

Sanderson*—would argue from this a homology between the structures standing in the most direct relation to these processes—that is, between the foliaceous axis of the moss and the prothallium of the fern, as the parts which bear the sexual elements ; and again, between the leaf-bearing stem of the fern and the theca of the moss, as immediately derived from the impregnated germ, and as originating the succeeding phase by a gemmiparous process (the formation of spores). Without at present entering into any arguments of a general nature on the correctness of this view, it may be simply remarked that the *primâ facie* aspect suggests rather a homology between the leaf-bearing stems in the two orders, in which case we should ascribe the different relations between them and the reproductive organs to the interposition of a process of gemmation at different periods of the genetic cycle—that is, just *before* impregnation in the fern, and immediately *after* it in the moss. Not that the resemblance is *merely* of a *primâ facie* kind. In some respects it grows upon us the more we contemplate it, for the persistent character of the leafy axis of the moss, and its frequently yielding many successive sets of sporiferous capsules, assimilate it, independently of structural features, rather to the stem of the fern than to its prothallium, which is an organ even more evanescent than the capsule of the moss, its existence terminating when the embryo formed in it has begun to germinate.† In this view the gamomorphic stage, which is not represented in mosses farther than by the formation of antheridia and archegonia as mere appendages of the typical axis, embraces in ferns and *Equiseta* the development and dispersion of spores, and their germination into prothallia bearing the sexual organs, either in different phytoids, as is generally

---

* Veget. Ovum. in Cyc. Anat. and Phys., Vol. IV.
† Jenner in Edin."Philos. Journal, N.S. III. 269. Radlkofer, Annals Nat. Hist. 2d Ser. XX. pp. 241, 344, 439.

the case in the latter order, or in the same one, as is the common rule in ferns. On the other hand, the protomorphic stage, which is so much protracted in Hepaticæ, and still more in mosses, is represented in ferns only by the confervoid chain of cells, termed the suspensor, which is the first result of the development of the impregnated germ, and at the farther end of which the embryo is formed.

In both groups the impregnated archegonial cell germinates while still within its receptacle, so that the resulting growth has the appearance of being simply an expansion of the structure immediately preceding ; thus the theca of the moss simulates a mere out-growth from the parent axis, while the prothallium, which is also really the parent of the fern-stem, has been described as its primordial leaf or cotyledon, from its resemblance to that organ in the higher plants ;* for as the embryonic structure advances, it sends downwards rootlets into the earth, and upwards a fern stem—the centre of growth still remaining in the prothallial frond.

The continuity of the organization is not broken in upon by any act of the nature of " seeding," till the dispersion of the spores. These particles, though physiologically equivalent in the two groups—inasmuch as in both they are asexual gemmæ, and not the direct products of impregnation—thus come to have very different homologies ; for they belong in one case to the protomorphic stage, and germinate more or less directly into the leafy axis, while in the other they are gamomorphic gemmæ from that axis, and develope the prothallium or sexual phytoid.

On the assumption of the line of homology now indicated, we may establish a parallel between the separate fern-fronds and the leafy shoots of the moss—observing that in both orders alike, free gemmæ are occasionally detached for the

* Carpenter Compar. Phys. 4th Ed., p. 507, Note.

multiplication of the plant—and we may farther compare the sporangia of the former to the perichætial involucre of the latter, both being modifications of the prevailing type of leaf-development. This is obvious in the moss, and it is indicated also in ferns by the circinate development both of fronds and sporangia, as well as by the occurrence of certain transitional forms, like those through which the ordinary leaves of *Blechnum* and *Osmunda* are depauperated into the spikes of fructification, or like the leaf-shoots which in some of the "viviparous" species take the place of the sporangia themselves.* In *Equisetum*, too, we can trace in the concentric arrangement both of the sporangia and their peltate receptacles, a marked relation to the general type of the whorls of primary and secondary branches, which represent the leaves—a relation of the same general character as that already noticed in the organs of *Chara*.

§ 8. LYCOPODIACEÆ AND RHIZOCARPEÆ.

The Lycopodiaceæ of the present epoch are a group of plants intermediate in habit between mosses and ferns. They resemble the former in their prostrate and ramose growth, and their minute scale-like leaves, while their affinity to ferns is indicated by the fibro-vascular system of the stem, and by the spores of the cauline fructification not being the result of any prior act of impregnation. In their process of reproduction they depart widely from mosses, and occupy a position in some respects intermediate between ferns and phanerogamic plants. But the process has not been satisfactorily traced out, except in *Selaginella* and *Isoetes*, which differ from the rest of the order in having two kinds of spores, a lesser and greater, termed respectively *microspores* and *macrospores*. The

* Lindley Veget. Kingdom, p. 75.

former are antheridia, containing cellules in the usual way, each with an antherozoid in its interior ; but these are not developed for a considerable time after the shedding of the spore. In about the same time the *macrospore* developes in its interior a prothallial layer, containing archegonia, in one of which the central corpuscule becomes converted into an embryo, the access of the antherozoids for impregnation being effected by the rupture of the spore-coats. In its small dimensions and internal position this prothallial layer presents an evident approximation to a structure presently to be noticed in certain phanerogamic ovules.

The first development of the impregnated germ is into the cellular suspensor. This grows downwards through the substance of the prothallium into the general cellular contents of the spore, and at its farther end the proper embryo is formed. By a change in the direction of growth the latter emerges from the spore, sending upwards a leafy axis, and emitting radicles from below, much in the manner of the embryo of a true seed. The axis increases by lateral pullulations, all assuming a prostrate position, and emitting rootlets as they creep along. Eventually they become independent phytoids by the decay of the first formed portion of the stem. The fructification is generally in distinct shoots, which assume the form of vertical spikes, with the leaves more closely imbricated, and bearing the sporangia in their axils.

In the small order of Marsileaceæ or Rhizocarpeæ, containing a few aquatic plants, the general character of the reproductive process is much the same as in Lycopodiaceæ. The two kinds of spores are contained in separate capsules within the same multilocular receptacle, termed a *sporocarp*. The process of germination comes still nearer to that of Phanerogamia than in the case just described, for provision is made for the entrance of the antherozoids, and the exit of the embryonic axis, not by the simple rupture of the spore-coats, but by a regular aperture or *micropyle*, evidently corresponding with the perforation so named in the coats of the ovule. No such aperture is required in the lower orders, where the archegonia are either naked as in mosses, or merely sunk in the surface of an exposed prothallium as in ferns. The pore going by this name in the sporangial cell of an Alga corresponds rather to the archegonial canal, than to the micropyle of these spores, or of true seeds.*

§ 9. REPRODUCTION IN THE GYMNOSPERMOUS PHANEROGAMIA.

The group of plants now known as Gymnospermeæ, including the Coniferæ and their allies, agree with the other Phanerogamia in the general character of the reproductive process—that is to say, the ovules which contain the germinal bodies are formed and matured in connection with the parent plant, and are fertilized by the action of pollen grains produced in anthers, which grow on the same or on different individuals, according as the monœcious or diœcious arrangement prevails in the species. At the same time they present some interesting peculiarities, which indicate an approximation to the process as performed in the higher Cryptogamia. The characteristic feature of the fructification is that the carpels which bear the ovules are not folded so as to enclose them, but merely bear them on their mar-

---

* The first discoverers of the prothallial structures of the higher Cryptogamia were Nägeli, Munter, and Suminski. A notice of the later writers is given by Professor Henfrey in the Annals of Natural History, 2d Ser., Vol. IX., p. 411. Hofmeister is the best known. An English edition of his works was announced some years ago, but has not yet appeared; it is understood, however, that the Ray Society has it in contemplation. A general survey of the reproduction both of Cryptogamic and Phanerogamic plants is given by Dr. Sanderson in the Cyclopæd a of Anatomy and Physiology, Art. "Vegetable Ovum," Vol. IV., p. 211. (Part XLV. March, 1855.)

gins, so as to leave them exposed, and allow impregnation to be effected by the direct contact of the pollen grain with the aperture of the coats at the apex of the nucleus, known as the *micropyle*.

The two coats and the nucleus of the ovule evidently correspond to the double-wall and cellular contents of the spore of the Rhizocarps, but it is in the development of a peculiar intra-ovular structure—the *Albuminous body*—that the principal interest of the comparison lies. It originates as a simple cell, but as it enlarges it generates in its interior a cellular mass, which may be held to represent the prothallium of the Cryptogamic spore. In its substance are subsequently developed two or three minute capsular bodies—the *Corpuscula* of Brown—each surmounted by a crown of four cells, like those forming the neck of the archegonium of the spore. The Corpuscula contain minute cell-like bodies, the one at the base corresponding to the germ cell of the archegonium, as well as to the "germinal" or "embryonal" vesicles of other ovules. Concomitant with this are the changes affecting the pollen grains, which, though they present certain peculiarities in Gymnospermeæ, are formed, as in other Phanerogamia, within a modified leaf or anther, by the quaternary division of the contents of the central cells of the parenchyma. From the exposed condition of the ovule in the Coniferæ, the grain finds direct axis to the micropyle, but it lies there for some time before it protrudes the pollen tube.

Schacht has observed that this outgrowth is connected with a process of cell-development in the interior of the grain, though he is not inclined on this account to allow any analogy with the antheridial cellules of the Cryptogamia. In the pollen tube he admits only the presence of free protoplasmic particles and starch grains.* Other

---

* Vegetable Anat. Phys. (Currey), p. 172-186.

authors, however, assert the existence of free cells in this situation, at least in certain families, as the Cupressineæ.* Hofmeister in particular mentions the formation of cells in the dilated extremity of the pollen tube, by a process of endogenous multiplication. The last formed cells contain granules, along with minute vesicles, which may be the ultimate stage of the granules, and fusiform particles, which are perhaps originally the nuclei of the vesicles, and which somewhat resemble the spermatia of Lichens and Fungi.†

The Pollen-tube, after insinuating itself into the tissue of the nucleus, has its growth arrested in some species for a whole season, and only recommences its progress when the "Corpuscula" are fully matured. It then penetrates through the overlying stratum of the nucleus, and through the wall and upper stratum of the albuminous body. Impinging finally on the summit of one of the Corpuscula, it displaces the rosette of cells here situated, and either sends down a process into the Corpusculum by invaginating its wall, or becomes merely flattened out over its summit, no aperture being formed either in the wall of the cavity or in that of the pollen tube. After this the germ cell at the base of the Corpusculum resolves itself, by a process of division and sub-division, into a group of eight cells, the four lower being the rudimentary embryos—of which, however, all but one abort—while the four upper become elongated, by a continuance of the sub-division, into so many cellular filaments or suspensors, whose growth pushes down the embryos—as in the spore of the Rhizocarps—into the underlying stratum of the nucleus.

This development in the ovule, the whole of which takes place while it is still attached to the tree, occupies a very long time, many Coniferæ not ripening their seeds till the next year after flowering. The fall of the seed arrests the

---

* Micrographic Dictionary, p. 518.
† Annals of Nat. History, 2d Ser., XIV., 427.

development, but it is renewed in germination by the upward growth of one of the embryos, and the consequent emergence of the plumule and radicle of the young plant, as in the case of other seeds.

This sketch of the reproductive process in the Coniferæ will suffice to show its leading relations, both to the other Phanerogamic orders and to the Cryptogamia.

The main points of agreement with the latter are :—

1. The nakedness of the ovule at the time of impregnation.
2. The development in its interior of the albuminous body and corpuscula, representing the prothallium and archegonia.
3. A certain amount of cellular development inside the pollen grain.
4. The arrest in the development of the pollen tube, which divides it into two stages, admitting of comparison with the maturation and emission of the microspore, and the subsequent evolution and discharge of antherozoids from it.

The points again in which the Coniferæ depart from all Cryptogamic, and agree with other Phanerogamic plants, are principally the following :—

1. The adhesion of the ovule to the parent plant till after impregnation and the formation of the embryo.
2. The lapse of time between the formation of the embryo and its evolution in germination; this is the period of the latent life of the seeds, during which their fall and dispersion take place.
3. The substitution of fovilla in a pollen tube, in

the place of free antherozoids, as the medium of impregnation.

The first of these points of difference is that which gives its most *obvious* character to the reproductive process—just as among Cryptogamia the feature most obviously distinguishing the Rhizocarps is the absence of an *external* prothallium, which is tantamount to the suppression of one of the two germinations *(prothallial* and *embryonic)* undergone by the fern spore. This maturation of the embryo *in situ,* of course, affects the position of the breach of continuity of organization, or the *dissemination* of the plant— the severance of the new generation from the old. It transfers it from the place occupied in the higher Cryptogamic orders—the early part of the gamomorphic stage— to the interval between the protomorphic and orthomorphic stages; or rather to the commencement of the latter, for the embryo is already fashioned in most species before the seed is shed. But in the Coniferæ, besides this, which may be called the great breach of continuity, there is a lesser one—the shedding of the pollen, which affects only the male element. Here, as elsewhere, this, of course, occurs prior to impregnation.

§ 10. REPRODUCTION IN THE ANGIOSPERMOUS PHANEROGAMIA.

The vast majority of the plants of the present epoch belong to this group, and principally to its higher or dicotyledonous division. The reproduction is throughout by pollen grains and ovules, but both, and particularly the latter, differ considerably from those of the Gymnosperms.

The carpellary leaf, instead of merely supporting the ovule, is wrapped round it to form a germen, generally of a more or less flask-shaped figure, with a neck or style resulting from the rolling up of its distal portion, and terminating in a stigmatic point denuded of cuticle. All direct contact of the ovule and pollen grain is thus prevented, and impregnation is effected by the grains adhering to the stigma, and from that sending down their tubes, through the lax tissue of the canal of the style, into the micropyles of the ovules in the germen below. No process of cell-formation has been observed in the interior of the pollen grain, and it is probable that its first protrusion is due simply to an endosmotic action, causing the contents—ensheathed in the extensible inner wall—to be protruded in finger-like processes through perforations at points where the outer cellulose coat has given way. But the farther advance of the tube must be effected by a proper growth, for its extension soon comes many times to exceed the size of the grain from which it was emitted. The ovule consists at first of a cellular nucleus, round which a double coat grows up, leaving only at the apex the pore termed the micropyle, and within which a cell cavity or embryo-sac is afterwards formed. This sac acquires much greater size in some cases than in others, and occasionally protrudes from the micropyle in the form of an "ovule-tube.[*] It contains only a semifluid granular matter, or at most a mass of very delicate cells. Among the contents, however, are generally seen two or three particles more conspicuous than the others, which have received the name of "germinal vesicles," but which some observers consider to be mere unwalled masses of protoplasm.[†] Unlike the germ cell of the Coniferæ, which lies at the bottom of the "Corpusculum," they occupy the space near the apex of the embryo-sac. The precise relations between the point of the pollen-tube and the embryo-sac were for some time the subjects of much discussion. Schleiden, Geleznoff, De Bary, Wydler, Schacht, and Tulasne, maintaining either an introversion or perfora-

---

[*] Dr. Dickie, Ann. Nat. Hist., N.S., I., 260.
[†] Griffith & Henfrey, Micr. Dict., p. 482.

tion of the sac, and the formation of the embryo from a process of cell-formation in the extremity of the pollen tube, while other botanists of note did not admit more than an intimate contact between the extremity of the tube and that of the embryo-sac.

The controversy may be said to be now mere matter of history, the former view having latterly been abandoned by its most distinguished supporters—Tulasne, Schleiden, and Schacht.* It seems to have arisen from the suspensor of the embryo having been mistaken for, or confounded with, the extremity of the pollen-tube. As it would appear, therefore, that there is no perforation either of the embryo-sac or of the pollen-tube, such commixture of their contents, as may be necessary for impregnation, must be held to depend on the transudation of the fovilla through the interposed membranes into the sac, in which it is now generally admitted that the embryo is developed, out of one of the contained "germinal vesicles." Impregnation through interposed membranes is certainly not according to the general analogy of the reproductive process, but it were premature to assert that it does not occur in other divisions of organic nature.† The "germinal vesicle" which has undergone impregnation becomes resolved, as in the Coniferæ, into two cells, the upper forming the confervoid suspensor, the lower the embryo.‡ The latter is at first a globular mass of cells, but generally, while the seed is ripening, cer-

---

* Henfrey—Annals Nat. Hist., 2d Ser., XVII., 343.

† In Dr. Carter's Observations on *Œdogonium*, before noticed, (ch. II., § 4), no bodily penetration of the antherozoids was visible. They seemed to degenerate into drops of reddish mucilage on the mucus-layer of the sporangium, and to be absorbed by a sort of endosmose. Annals of Nat. Hist., 2d Ser., XVIII., 81.

‡ In the phanerogamic ovule, the suspensor is generally formed after the development has commenced of the proper embryo; its prior formation in the Coniferæ is one of the points in which this group presents a transition to the characters of the Cryptogamia.

tain rudimentary organs are formed—the plumule, radicle, and cotyledons, representing the axis, root, and leaves of the future plant—and it is on differences in this respect that the great division into Monocotyledons and Dicotyledons is based.*

The whole process of development is more rapid than in the Coniferæ, there being no arrest in the growth of the pollen-tube, and the flowering and seeding of the plant being generally accomplished in the same season ; but there is an entire agreement in the main feature—the adhesion of the ovule to the parent plant till the maturation of the embryo. In some exceptional cases the connection is not entirely broken off even then ; the inextricable thickets of Mangrove, with which swampy tropical shores are fringed, are said to be due to the property which the seeds of this tree have of germinating while still attached to the branch.†

From this sketch of the reproductive process, it appears that the Angiospermous Phanerogamia differ both from Coniferæ and from the higher Cryptogamia (Rhizocarps, Ferns, &c)—firstly, in the absence of any accessory cells at the summit of the embryo-sac, like those which form the crown of the corpusculum and the styloid neck of the archegonium ; secondly, in the non-development of any distinct tissue in the ovule, like the albuminous body or the prothallium ; in compensation for which, as it were, we have, thirdly, an additional outer envelope—the germen or ovary. From the Cryptogamic orders they differ farther in the separation, as it would seem, of the two elements by the continuous membranes of the pollen-tube and embryo-sac. In Phanerogamia we have two distinct genital canals, but neither of them corresponds to the archegonial canal

---

* In the Orobrancheæ and Orchidaceæ, the embryo reaches no higher development, in the ripening of the seed, than a globular mass of cells.

† In apples, &c., the seeds are sometimes seen in a state of germination.

of the fern ; one of them—the micropyle—leading through the coats of the ovule, appears first when the prothallium becomes internal, as in the Rhizocarp ; the other, that of the style, being entirely extra-ovular, can, of course, have no existence when the spores are deciduous, as in the Cryptogamia, nor even when the ovule is uncovered, as in the Coniferæ.

The true homology of the relation in which the various intra-ovular structures of the Gymnospermeæ stand to those of other Phanerogamia, is a point on which authors are either not agreed, or else are so loose in their terminology as to obscure their real meaning,* but the bearing of the several envelopes to each other, and to the germ which they enfold, may be represented to a certain extent, as follows, both in these groups and in the higher Cryptogamia, without entering into the discussion :—

>Angiosperms have three germ-envelopes—viz.,
>Germen—pervious through the style ;
>Ovule-coats—pervious by the micropyle ;
>Embryo-sac—imperforate.

Gymnosperms have also three germ-envelopes, though not all homologous with the foregoing, viz. :—

>Ovule-coats—pervious by the micropyle ;
>Albuminous body (prothallium)—perforated by the pollen-tube ;

* Thus the term *endosperm* is applied both to the tissue of the albuminous body of the Coniferæ, and to the cellular growth which takes place *within* the embryo-sac in the course of development in many plants, and which remains as a permanent constituent of the seed in Ranunculaceæ and Nymphæaceæ ; but, of course, the identification of these structures is incompatible with a homology between the embryo-sac and the corpusculum or archegonium—as much so, indeed, as the notion implied in the old name of *Pistillidium* applied to the last mentioned organ. See the article on Vegetable Reproduction in the Cyclopædia of Anatomy and Physiology.

Corpusculum (archegonium)—imperforate ;
Rhizocarps have also three germ-envelopes ;
 Spore-coats—pervious by the micropyle ;
 Prothallium ⎱
 Archegonium ⎰ pervious by the canal of the latter.

In Ferns, from the spore being merely a provisional structure, and from the prothallium being exposed, and not enclosing, but only supporting the archegonium, the latter is the only germ-envelope, and its canal the only passage of access.

In mosses, as has been shown, there is no body corresponding to the spore of the ferns ; both it and the derivative prothallium, which in that group had already ceased to be germ-envelopes, have now no longer any existence, the archegonia being attached directly to the leafy axis of the plant.

§ 11. The general conclusion to be drawn from the foregoing survey of the leading modifications of the reproductive process in plants, appears to be that the difference between the higher and lower species consists mainly in the constructive energy of the former being as it were concentrated on that embryogenetic development, whereby a higher degree of organization is attained in the typical or orthomorphic condition ; the other two stages in the cycle of propagation being represented only by the maturation of the floral organs, and the formation of the cellular mass which is the earliest condition of the embryo. The latter process, though really the beginning of a new cycle, is so blended with the concluding or gamomorphic stage of the preceding, and both are so merged in the following orthomorphic phase of development, that their individuality is quite lost, and they appear as mere subsidiary processes, affecting only certain organs of the typical plant, whose term of life is popularly supposed to commence from the germination of the seed, though this is really nothing more than the *evolution* of an embryo already elaborated. As we descend to lower forms, we find one or other of these subsidiary stages becoming more prominent by the diffusion over them, as it were, of plastic energy, abstracted from the elaboration of the typical form, which declines proportionally in complexity of organization. In the Gymnospermeæ and the higher Cryptogamia it is the gamomorphic stage which gradually acquires importance, the protomorphic not being conspicuously brought out till we reach the mosses and lower orders.

In the following chapter a similar survey will be taken of the modifications of the process in the animal kingdom.

III.

SURVEY OF THE REPRODUCTIVE PROCESS IN THE ANIMAL KINGDOM.

§ 1. THE function of reproduction in the Animal Kingdom, while it embraces wide variations in accessory points, presents a great, if not an absolute, uniformity in the production and conformation of the sexual elements. The spermatic particles, or spermatozoa, are developed as the solitary nuclei of secondary cells—the vesicles of evolution—which are in turn generated in variable numbers within the cells occupying the cavities of the spermatic gland. When liberated by the rupture of their envelopes, the particles float freely in the fluid secreted by the gland—singly or in bundles as the case may be. Their normal form is that of a minute rounded body, with a long filiform appendage or *cilium*, by the vibration of which an onward motion is given to the whole corpuscule. Both parts are sufficiently apparent in the higher animals, but among the lower tribes, as, for instance, in insects, we find the body sometimes so much attenuated as to be undistinguishable from the cilium, while in other cases the latter appendage disappears, and with it the motile power. This modification obtains in some Crustacea and Entozoa.*

The germinal corpuscula, on the other hand, appears in the form of a minute nucleated body, and is known as the *germinal vesicle*. Like the spermatozoon it seems to be developed as the nucleus of a secondary cell (the *ovum*), which is generated in the interior of another (the *ovisac*),

* Siebold's Compar. Anat., § 348, 290, 117.

lying in the substance of the ovarian gland—such at least is the case in the higher animals; but it is very doubtful if these relations can be satisfactorily made out in all the lower orders, though the ovum has very constantly the character of a nucleated cell. It is commonly only the outer envelope that dehisces, the ovum itself being merely penetrated by the spermatic secretion, for the accomplishment of the act of fecundation. It is not yet absolutely certain that the spermatozoa always penetrate into the ovum in form; possibly in some cases there may be nothing more than a transudation of their liquified substance. Their bodily presence within the ovum, however, has now been so frequently detected as to afford ground for believing a formal penetration to be the usual rule.* We find, indeed, what may be considered as a special provision for their introduction, in the *micropyle*, which is found in many eggs, and consists of one or more apertures at one of the extremities. The functional import of the micropyle, as a passage of access for the spermatozoa, is rather confirmed by the circumstance, that its homological relations are not the same in all ova. In such as have been originally attached by a

* The first distinct observations of spermatozoa in the ovum appear to be those of Dr. Martin Barry on the rabbit (Philos. Transact. 1840, 1841, 1843, March and June. Edin. Philosoph. Journal, Vol. LV., 326, LVI., 36). They have also been observed by Dr. Farre in the earthworm, by Dr. Nelson and Meissner in *Ascaris* (Philos. Transac. II., 1852), and by Dr. Newport in the frog (Philosoph. Transactions, 1853, p. 266-281). In the following year Bischoff, who had before discredited the penetration of the spermatic particles, himself observed it, and since then Siebold has detected it in the ova of the bee (True Parthenogenesis, p. 85), and Gengenbauer in those of Hydrozoa (Huxley, Oceanic Hydrozoa, p. 22). Claims have also been advanced for Prevost and Dumas, Wagner and Keber. Keber's statements have not received much consideration from naturalists generally, except as regards the discovery of a micropyle in the ova of *Unio* and *Anodonta*. A notice of the successive discoveries in this department is given by M. Claparede—Annals of Nat. Hist., 2d Ser., Vol. XVII., 298.

stalk—as in those of Echinodermata and Helmintha—the portion of the hollow pedicle which remains in connection with the ovum becomes the micropyle, but in other cases a special aperture appears to be formed for the purpose, in the previously continuous wall of the ovum. A micropyle has now been observed in the ova of all insects, and in those of Acephalous Mollusca, and of Osseous Fishes, as well as in those of some Crustacea, Annelida, Echinodermata, and Nematoid Entozoa. In the case of the Cephalopoda and Batrachia the evidence is less satisfactory, and no trace of it has yet been detected in the Mammalian ovum, though the presence of spermatozoa in its interior has now been verified by more than one observer. Here, therefore, we can only conjecture that they may gain access by a sort of boring action, or by the formation of extemporaneous apertures.* The micropyle is certainly not the only provision for the purpose. In some cases the proper wall of the ovum disappears before it comes into connection with the Spermatozoa, which then penetrate into it over its whole extent, their filamentous extremities giving the surface a ciliated appearance. This has been observed by Meisner in the *Lumbricus*, and may perhaps occur also in some Hirudinei and Mollusca. In the Helmintha again, the spermatozoa are frequently brought in contact with the ova while the latter are yet in process of formation, as will be more fully mentioned in the notice of that class. In the Hydrozoa, too, impregnation seems to be effected in the nascent condition, as it were, of the ova, for these bodies do not appear to have any proper investing membrane,

---

* In the ovum of the rabbit Dr. M. Barry describes (Op. Cit.) the formation of a chink for the entrance of the Spermatozoa. In that of the frog, according to Mr Newport, there is no perforation, aperture, or fissure of any kind visible to the eye, but he has distinctly observed the penetration of the Spermatozoa in considerable numbers and with great rapidity (in less than a minute). Philos. Transactions. 1853.

when they are detached from their capsules and brought in contact with the spermatozoa.*

About the time of impregnation the germinal vesicle is generally stated to disappear. In the Batrachia Mr. Newport is inclined to think that it is ruptured by the pressure of a brood of minute cells formed in its interior, and its contents dispersed in the form of clear spherules through the yolk or mass of oleo-albuminous particles with which the ovum is filled. This may serve in some way as a preparation for fecundation; at all events he is positive that in the ova of the newt and frog the disappearance of the vesicle takes place before impregnation, and not in consequence of it.† The first obvious result of this act is the repeated cleavage or segmentation of the original contents of the ovum, and the formation thereby of a mass of cells or plastic spherules, out of which the embryo is developed either immediately, or with the intervention of some of those diverse forms which occur in cases of so-called Alternation of Generations.

After impregnation, and when the process of segmentation is about to begin, a clear nucleated cell—the " embryo cell"—is generally observed in the interior of the yolk. In

---

* Gegenbaur, quoted by Huxley, Oceanic Hydozoa (Ray Soc.), p. 22. These diversities in the mode of access of the Spermatozoa may be tabularly represented as follows:—
    By a micropyle, in Echinodermata, Worms, Insects, Crustacea, and Osseous Fishes (possibly also in some Reptiles *(Hyla)* ;
    By fissure of the wall of the ovum, in Mammalia; (?)
    By penetration through its substance, in Batrachia ;
    By commixture before the ovum is coated over, in some Trematoda and Nematoidea, and in Hydrozoa;
    By dissolution of the investing membrane, in *Lumbricus*, and perhaps in some Hirudinei and Mollusca.
The subject is noticed with some detail in Professor A. Thomson's article (Ovum) in the supplementary volume of the Cyclopædia of Anatomy and Physiology. Also in the paper by M. Claparede, quoted above.

† Philosoph. Transactions (1851), p. 169.

Entochoncha mirabilis, a molluscous animal, this, according to J. Muller, is identical with the germinal vesicle, which in that species never disappears. In other cases, as *Ascaris*, it has been supposed to arise from the nucleus of the vesicle. Some such connection has been assumed, indeed, even when the vesicle seems to disappear entirely, as it does in most ova. It has been thought that the deliquescence may not extend to its whole contents, and that the "embryo-cell" may originate from some residuary portions, as from some of the clear spherules, before referred to, in the ova of Birds and Batrachia. In the Hydrozoa, according to Gegenbauer, at least in the Corynidæ, Calycophoridæ, and Physophoridæ, "the germinal vesicle" does not disappear, but its division immediately precedes that of the yolk, so that its progeny must eventually become the "embryo-cells" of the division masses.* The observations of the same naturalist on *Sagitta*, and of Professor Huxley on *Pyrosoma*, also tend to show that the "embryo-cells" are the lineal descendants of the germinal vesicle.

The division of the "embryo-cell" immediately precedes that of the yolk. The segmentation is commonly effected by each division of the "embryo-cell" becoming coated over with a corresponding portion of the granular matter of the yolk, so that every one of the multitude of minute spherules, into which the latter is ultimately resolved, contains, as its nucleus, a derivative of the "embryo-cell." But in a few cases, confined, so far as is yet known, to some Nematoid and Cestoid worms, the progeny of the original "embryo-cell" do not coat themselves in this way, but seem rather to absorb the granular and fluid matter, growing, as it were, at its expense, so that the yolk entirely disappears by the time the segmentation is complete. Professor Huxley considers that a somewhat analogous pro-

---

* Huxley's Oceanic Hydrozoa, p. 22

cess of cell-formation occurs in the ova of *Pyrosoma* and *Salpa*.*

The embryo—whether the direct or indirect product of impregnation—differs in the majority of cases from the adult form, not only in size, but in many points also of structure and configuration, and the progressive changes which it undergoes before acquiring perfect conformity to the organization of the parent go under the name of *metamorphosis*. Sometimes they occur within the embryonic envelopes, but at other times the young is extruded while still in an imperfectly developed state. Such naked embryos are termed *larvæ*—masks, as it were, disguising what is ultimately found to be the true aspect of the species.

In some Invertebrata, Fishes, and Reptiles, and still more strikingly in Birds, the segmentation appears not to extend to the whole contents of the ovum, but the exception is more apparent than real, for the yolk of the fertilized egg in such cases contains part of the granular matter within the ovisac, over and above the proper substance of the

---

* Annals of Nat. History, 3d Ser. (Jany., 1860), p. 35. There is perhaps more force in Mr. Huxley's other suggestion of an analogy with the Bird's ovum, though in this view the part wanting would not be the true cleaving yolk, but the wall of the proper germinal vesicle, and the true ovum would be represented by what he calls the germinal vesicle, not by the unwalled contents of the ovisac, which would correspond rather to the adventitious or yolk-food of the Bird. The homologies would stand as follows :—

Bird.		Pyrosoma.
	Vitelline membrane...........wanting.	
	Food yolk......................liquifying yolk of Huxley.	
	Temporary zona of Meckel { wall of " germinal vesicle" of Huxley.	
	True or primitive central granular yolk .... } Contents of above.	
	Wall of germinal vesicle.......wanting.	
	Macula of do....................germinal spot.	

Reasons will be given afterwards for regarding the *wall* of the germinal vesicle as a non-essential structure, though one certainly of great constancy in the animal kingdom.

original ovum, owing to the obliteration of the primordial wall of the latter, and the subsequent formation of a new limiting membrane—the vitelline—nearer the wall of the ovisac.* The supplementary material does not undergo segmentation, and goes under the name of food-yolk ; it is evidently provided to secure a larger supply of nutriment within the egg, in cases where none is to be obtained from without during the development of the embryo. It is not present in the Mammalia, where the fœtus is nourished through the placenta, nor in cases among the lower orders in which the embryo is speedily set free as a larva.

It is to diversities in such secondary points as those last alluded to, that the infinite variety is due, which the Animal Kingdom presents in the details of the process of reproduction. In the points more essentially connected with the formation and fertilization of the germ, there is, as has been mentioned, an essential sameness throughout, for the phenomena which at first sight appear most eccentric —those connected with the alternation of generations—do not arise from anomalies in the sexual process itself, but from the interpolation of an independent process of Monogenesis, at different periods in the life history of the species.

It is the relation between these two associated processes which will form the principal subject of attention in this summary of the modifications of reproduction in the more important groups of animals ; the other peculiarities of the function will be alluded to only in so far as they help to illustrate this point.

§ 2. REPRODUCTION IN THE PROTOZOA.

Following the same ascending order as in the last chapter, we are met at once on the threshold of the Animal Kingdom with a form of life to which we are not yet in a

---

* Thomson in Cyc. Anat. and Phys., Part LIV., p. 77.

position to apply unreservedly any of the general principles above stated, as hitherto no indications of sex have been discovered in the majority of the species. Such points, however, will be stated as seem to have any distinct bearing on the subject before us.

Under the name of Protozoa are comprehended certain unicellular animals having more or less affinity with the infusorial animalcules, in the restricted sense in which this term is now used, after the elimination of the embryos of the higher species, of unicellular plants, such as have been already considered, and of other extraneous forms. But there can be little doubt that names still occur in the list of Protozoa, which do not represent real species, but only the embryonic condition of such as are referable in their adult state to far higher types of organization, for a transformation which has now been satisfactorily traced in so many instances, may fairly be suspected to occur in not a few others still referred to the lower group. The detection of such cases must be left to patient observation, with the conviction that, though this is a slow process, it will eventually yield results which may confidently be relied on.

But even in cases where long continued observation shows that the unicellular organisms in question never attain any higher type of structure, there still remains behind another difficulty—namely, to determine whether they are of an animal or vegetable nature, and the grounds on which this is to be decided, or the points of distinction between the Protophyta and the Protozoa, form a question about which naturalists are not yet quite in agreement, though sensibly approaching to it. The following are the points now generally admitted as distinguishing the unicellular animal from the vegetable :—

1. Contractility of the substance or of the bounding wall of the organism, which is a main agent in its locomotion, though ciliary action is also employed, and to a much

greater extent than in the vegetable cell, whose motile powers are consequently less marked and less constant.

2. The predominance of albuminous compounds over those ternary principles, such as cellulose, starch, and chlorophyll, which form the bulk of the vegetable cell and its contents, and the absence of the green colour which the action of light on these vegetable principles evolves.

The Protozoa, on the elimination of these spurious or intruded species, constitute a truly natural group, but one, at the same time, hardly admitting of any good general definition, applicable to the several forms it includes, being distinguished from those referable to the other primary divisions of the Animal Kingdom chiefly by negative characters, such as the absence of a nervous system and of organs of sense, and in many even of a distinct alimentary apparatus. They are sometimes described as unicellular, with a *nucleus* or minute solid particle, and certain clear spaces or vacuoles in their interior,* but it is to be observed that no true cell-wall is developed in the lowest forms—the body consisting of a mere mass of plastic jelly, without any distinct membrane bounding its exterior.

In the greater number of the Protozoa, reproduction is only as yet known to take place in the way of monogenesis. Two leading modifications of this are admitted, though they are not always to be distinguished from each other—*fission* and *gemmation*.† Fission, or the spontaneous separation of the body into two or more segments, prevails mostly in the higher group known as Infusorial Animalcules—gem-

---

* Greene's Manual of the Protozoa, p. 2.

† Fission, as Professor Owen remarks (Parthenogenesis, p. 10), though it presents a wide *prima facie* diversity from ordinary gemmation, in half of the body of the parent, instead of only a small portion of it, going to form that of the offspring, is after all only a modification of the other process, the difference depending on the very small size in gemmation of the portion of parenchyma which takes on the new development, in relation to the whole body, and on its superficial position.

mation, or the development of external buds, in the Rizopods and Sponges. In the latter case there is a tendency—though not so general as in the vegetable kingdom — for the gemmæ to remain in adhesion with the basis from which they have been budded off, so as to give rise to composite multicellular organisms, whose configuration depends on corresponding variations in the mode of gemmation.

Of these aggregated Protozoan structures, the Foraminifera, and especially the Sponges, are the forms that attain the largest dimensions.

Indications, however, are not wanting of the occurrence of sexual reproduction, though as yet they have been recognized only in a few isolated cases. In Tethya, an animal of this division, ova and spermatozoa have been detected by Mr. Huxley.* They do not appear to be formed in special organs, but occur in mixed masses within the spicular investment of the common organism. The other cases are those observed by Balbiani among the Infusoria, as described in a late communication to the Academy of Sciences of Paris. This author has met with indications of the process in six or seven species representing different groups, but confines himself in this paper to a description of it as it occurs in *Paramœcium Bursaria*, the species in which he has been able to trace it most completely.† For several generations the *Paramœcia* multiply by spontaneous fission, each of the two new individuals obtaining half the primitive nucleus, but under the influence of conditions, of which we are still ignorant, the species propagates itself by sexual concourse. When the period for this arrives, the individuals are found coupled together in pairs, adherent laterally, and, as it were, locked together, with the similar extremities turned in the

---

* Annals of Nat. Hist., 2d Ser., VII., 370. Also by Lieberkühn in *Spongilla*, Op. Cit., XVII., 412.

† Annals of Natural History, 3d Ser., I., 435. (Comptes Rendus, March 27, 1858, p. 628). Greene— Manual of Protozoa, p. 72.

same direction, and the two mouths closely applied to each other. In this state the two conjugated individuals continue moving with agility in the liquid, and turning constantly round their axis.* There is nothing before copulation to announce the evolution of the sexual elements; it is during the act itself, of which the duration is prolonged for five or six days, or even longer, that their development takes place, at the expense of the nucleus and nucleolus of each animalcule. The nucleolus undergoes a considerable increase in size, and becomes converted by sub-division into two or more capsules, which contain ultimately minute fusiform bodies with fine filamentous extremities, regarded by Balbiani as spermatozoa. The corpuscules, which he considers as of a germinal nature, are certain spheroidal bodies, with indistinct central spots, formed in the substance of the nucleus with or without its previous segmentation. Fecundation seems to be effected by a transference of one or more of the spermatic capsules, through the closely adpressed mouths, from the body of one animalcule into that of the other. They continue to increase in size after their transference has taken place, one only arriving at maturity at a time. Five or six days after copulation, minute rounded germs make their appearance, and in the course of development are extruded from the body of the parent animal, but for some time after this they remain adherent to its exterior by means of the knobbed tentacles or suckers with which they are provided. At length they detach themselves, lose their suckers, acquire a mouth in their stead, and, becoming furnished with vibratile cilia, take on the aspect of adult *Paramæcia.*†

---

* In a second communication Mr. Balbiani generalizes his conclusions, and extends them to several other Infusoria, as *Euplotes*, *Chilodon*, *Spirostomum*, and *Orytrichina*. Annals of Nat. Hist., 3d Ser., II., 410.

† Dr. A. Thomson in Cyclop. Anat. and Physiol., Art. *Ovum*, p. 7. Greene's Protozoa, p. 51-71.

A sort of conjugation has been observed in some Infusorians, and particularly in the allied order of Gregarinidæ, consisting mostly of parasites infesting the intestines of insects ; but that attempts to assimilate it to the conjugation of the Protophyta are at least premature, is shown by the diversity of the two processes in some important points. For, firstly, the fusion which has been observed in *Vorticella* and *Gregarina* by Stein and others, is not always confined to a pair, but occasionally three have been seen to coalesce in this way ; and, secondly, the union does not seem to be of the same nature, for after coalescing in this manner, the animals have been seen to separate again in all their integrity. Where it really is preliminary to the formation of embryos, we may, therefore, more reasonably consider it as a kind of intimate copulation, like that which has just been described from Balbiani.*

In connection with the reproduction of the Infusoria, two other points seem to call for a short notice—encystment, and the transformations which have been described by some authors.

The encysting process has been most accurately observed in *Vorticella*, but it probably occurs in all animals of the class. A *Vorticella* about to become encysted contracts slightly, closes the peristome or ciliated depression in which the mouth is situated, and envelopes itself with a mucous

* In some of the species observed by Balbiani, the union amounted to an actual fusion of the individuals for more than two-thirds of their anterior part, and in all probability many supposed cases of incipient fission have really been of this nature. Among the Entozoa, still more striking cases of so-called conjugation have been observed. In *Syngamus* there appears to occur a real fusion of tissue, and in *Diporpa* even a coalescence of certain viscera between the two individuals, which, indeed, were first described as one duplicated animal ; yet, in its bearing on reproduction, the act seems more allied to ordinary copulation than to conjugation as it occurs in the Protophyta, as both the individuals concerned have proper generative organs of the ordinary type of the class. See Annals of Nat. Hist., 2d Ser., VII., 428.

secretion which gradually consolidates into a hard shell. In some cases the encystment seems to depend on exposure to cold or drought, and is then probably simply a means of protection from these influences—a sort of hybernation—for the animal may remain unchanged in the cyst, and, when humidity and warmth are restored, it may burst its envelope and resume its former life. But in other instances it appears to have a physiological import, and to be preliminary to certain transformations of the animalcule itself. Stein mentions several metamorphosis undergone by the Vorticella in the interior of the cyst, especially its breaking up into a number of minute corpuscules, which, on reaching maturity, are shed by the dehiscence of the shell, and serve as free gemmæ for the multiplication of the species. He describes also the conversion of the cyst itself into an *Acineta*, by the protrusion from its exterior of the characteristic knobbed tentacles. The convertibility of these two forms, and many others of Stein's conclusions, are rejected by Lachmann on the ground that sufficient care was not taken in the isolation of the specimens observed. More lately, however, M. d'Udekem has reasserted the derivation of acineta forms from the encysted *Vorticella*, though he differs from M. Stein in his account of the metamorphosis. He describes the transformation of the *Vorticella* within the cyst into a simple ciliated Infusorian *(Opalina)*, by its dissolution into granules, the exterior layers of which coalesce to form an integument, by a process somewhat resembling the formation of the blastodermic membrane of the ovum. The ciliated body after escaping from the cyst, which is ruptured by its growth, is transformed into an *Acineta*. By careful isolation, and the observation of intermediate forms, M. d'Udekem has satisfied himself of the reality of this change. Ciliated embryos, formed from the nucleus, have been seen by many observers to be discharged from the *Acineta*. Stein observed these, though he was

never able to trace their farther progress, but he conjectured
that they might give origin to Vorticella forms, to complete
the genetic cycle. It is possible this may be the destiny of
those acineta-embryos which M. d'Udekem has seen to become encysted, but it is now clearly ascertained, both by
this observer and by J. Müller and Lachmann, that they
may also be directly developed again into fresh *Acinetæ*.*

§ 3. REPRODUCTION IN THE COELENTERATA.

Until of late the lowest of the four primary divisions of
the Animal Kingdom admitted by Cuvier—the *Radiata*—
remained much in the condition in which that illustrious
naturalist found the whole invertebrate series—a sort of
lumber store, in which all forms not readily reducible
under the three higher divisions were conveniently stowed
away. Even after the labours of later authors had extri-
cated from the confused assemblage various aberrant forms
of the Molluscous and Articulate types—as by associating
the Polyzoa with the Tunicata, the Lernæadæ with the
Crustacea, and the Entozoa with other vermiform tribes—
it still remained as impossible as ever to establish any community of organization among the residuary species. Later
researches have shown, however, that the restricted Radiata
fall into three groups, all equally natural, though of very
different relative value—namely, the *Protozoa*, which have
just been noticed, the *Coelenterata* and the *Echinodermata*.
Of these the two former are now ranked as primary divisions, while the position of the last, which is evidently of a
subordinate character, still continues to be one of the great
puzzles of systematic zoology.

The Coelenterata, as established by Leuckart and Frey,
coincide with the group termed *Nematophora* by Professor

* Annals of Nat. Hist., 2d Ser., IX., 471, 3d Ser., IV., 1.

Huxley, and undoubtedly from a very natural assemblage, for in essentials a very complete unity of organization is to be traced through all the species. Along with this, however, there is so much variety in many adventitious points, affecting both the development of particular organs, and their mode of arrangement, that the general appearance is often strikingly unlike in the different sections of the group. The reproductive process seems to present a corresponding diversity, some species furnishing the most remarkable examples of alternation, while in others no phenomena of the kind have yet been noted. How far these variations may be brought within the scope of a common law will afterwards be considered ; at present it may be sufficient to remark that in the protomorphic or early development of the Cœlenterata, the principal point of interest is that the germinal mass becomes covered with a membrane bearing cilia, by whose play it moves freely through the water like an infusory animalcule. When the action of the cilia ceases, a mouth is formed on one side, and the embryo, assuming a cup-shaped form, becomes transformed into a polype.

Gemmation is generally a very conspicuous feature of the polypiform phase. It is by repeated pullulation of gemmæ, and their continued adhesion to the parent stock, that those composite structures are formed, so characteristic of the group, which have received the name of *zoophytes* or plant-animals, from their resemblance to ramose growths of a vegetable nature.

The polypiform condition is the most permanent stage of development, and for this and other reasons, which will afterwards be given, it is here considered as the orthomorphic or typical phase in the genetic cycle or life history of the species. In many cases, however, it is not that which matures the sexual organs. In the higher division, indeed, of the group (Actinozoa of Huxley) containing the Asteroid and Helianthoid polypes, and also in the common *Hydra*, the ova and spermatozoa are borne directly by the polypiform zoophytes into which the germ passes, by a continuous course of development,* but in the compound polypes allied to *Coryne* and *Sertularia*, and in the Calycophoridæ and the Physophoridæ of Huxley—all referable to his lower section of Hydrozoa—we have, as a very common arrangement, an alternation of forms, occurring in what has been distinguished as the Gamomorphic stage, *i.e.*, in connection with the evolution of the proper reproductive organs—the sexual elements being developed, not in organic union with what may be considered as the typical form, but in peculiar zooids detached from it. These go under the name of Medusoids, and though there are great diversities of detail, which will afterwards be adverted to, in their mode of origin from the polypiform zoophytes, and in their size and completeness of organization, there is a certain uniformity in their general structure, which consists of a natatory organ, developed round the spermatic or ovarian sac, and assuming the campanulate form, so prone to repeat itself at all points in this type of organization.

The sexual zooids of the compound polypes are of very minute size and rudimentary organization, while those of the Lucernarian section acquire a much larger size and more elaborate structure; so that although the two forms were associated in one order—as the naked and hood-eyed Medusæ—long before their true derivation was known, yet the idea of their being both alike the homologues of detached ovaries was long of making its way to the general acceptance it has now obtained, on account of the disparity between the structures themselves, and still more on account of the disproportion of their development to that of their

* The reproductive process seems to be direct also in the Ctenophora, from the few observations we have on the subject. See a notice of the development of *Cydippe*, in Edinb. New Philos. Journal, N.S., IV., p. 89, by Dr. T. S. Wright.

respective parent stocks; for while in the Hydroid and Sertularian Polypifera the zoophytic stem is a far more conspicuous object than the Medusoid, the polypoid stock, or larva, as it has been incorrectly termed, of the other form is quite an insignificant organism, in comparison of the colossal hood-eyed Medusæ which originate from it.

The Medusa-zooids seem to be almost universally of separate sexes, and most commonly it would appear that all those from one polype or polypidom are of the same sex—an arrangement comparable to that termed diœcious in botany.*

As the reproduction of the Hydrozoa will again come under review, in considering the nature and relations of the phenomena of alternation, it seems unnecessary at present to go into farther details on the subject.

§ 4. ECHINODERMATA.

This division of animals, so natural in itself, but so puzzling in its relations to other leading groups, presents us with phenomena in the protomorphic stage of development, not inferior in interest to the gamomorphic zooids of the class last under consideration. From the observations of various naturalists, and particularly of Prof. J. Müller, who has been the principal labourer in this field, it appears that the germ of an Echinoderm has at first the character of a ciliated Infusorian, with a tendency to a bilateral form. In some species of Starfish *(Echinaster, Asteracanthion)* the development of the typical Echinoderm takes place by a nearly direct process. The ciliated body soon manifests a distinction of parts, a four-lobed portion appearing at one extremity, by which the nascent animal adheres to any support; but this appendage is more than a mere pedicle of

* Dr. Wright in Edin. Philos. Journal, N.S., IV., 88.

attachment, as it has a mouth and gastric cavity of its own. The back of the ciliated body assumes a polygonal form, and is directly transformed into the starfish, which developes a new mouth in the centre of the ventral surface, a little to one side of the pedicle, the latter appendage gradually disappearing by a process of absorption.

But in the great majority of the class there is in the course of development a marked change, of the nature of alternation. In all these cases, a decided breach occurs in the continuity of the *process*, by the establishment of a new focus of organization ; and in not a few there is a breach of *structure* also, in the mechanical separation of the later from the pre-existing growth. The first steps of development—those which directly affect the germ—consist in the formation of an alimentary canal, with oral and anal openings, and in the disappearance of the cilia on the external surface, except in particular spots, especially along a circle surrounding the oral region. These are followed by certain alterations of external form and configuration, differing in the different sections of the class, and distinguished by particular names. That characteristic of the Holothuridæ has been termed *Auricularia*, and is of a cylindrical or barrel shape, girt with numerous ciliated rings, and not unlike the larva of some Annelida. That of the Asteridæ, known under the name of *Bipinnaria*, attains a considerable size— an inch or more in length—and assumes a very extraordinary form, from the development of the ciliated region in front of the mouth, which throws out several long processes on each side, and at the anterior extremity two fin-like expansions, placed one above the other. But the most elaborate specimen of such provisional organization is presented to us in the *Pluteus*—the form from which the Echinidæ and Ophiuridæ are derived. The general form of the Pluteus is that of a quadrilateral pyramid, dome-shaped above and slightly excavated at the base, the corners of which are prolonged into straight slender legs, strengthened by filiform rods of calcareous matter reaching to the summit of the dome. The mouth projects as a proboscis from the middle of the concave base, and the circle of cilia surrounding it fringes the circumference of the base of the pyramid, and the four projecting processes, which in swimming are directed forwards.

The new process of organization always originates in a diverticulum of the dorsal integument of the germ, which grows inward and lays the foundation of the future water-vascular system, on which the other organs of the Echinoderm are subsequently modelled.

In *Holothuria*, the new formation amalgamates to a great extent with the germ structures, and no part is absolutely cast off, though the original mouth is obliterated, and a new one formed on what was the dorsal aspect of the Auricularia. In the Echinidæ and Ophiuridæ, and also in the Asteridæ, a new mouth is formed much in the same way; but the only organ of the original appropriated by the Echinoderm is the alimentary canal. As development advances, a difference appears in the two cases, for while the unappropriated part of the Pluteus vanishes, the Bipinnaria retains its original form more persistently, and gradually becomes detached from the Starfish. The latter then sinks to the bottom, and creeps by its newly developed sucking feet, while the protomorphic zooid—eviscerated as it is by the process—still swims about for some time as before, but eventually perishes.*

No alternation of forms is known to occur in this class, after the acquisition of the typical organization, nor any phenomena at all of the nature of gemmation, except we are to consider the development of the sexual organs, or the regeneration of lost parts, as rudimentary manifestations

* An. Nat. Hist., 2d Ser., VIII., p. 4.

of a process of this kind.* Nor have we any remarkable structural metamorphosis in this stage, except in the pedunculated species, some of which at least are known afterwards to become free by separating from the calcareous stem. In this way the so-called *Pentacrinus Europeus* is converted into the *Comatula rosacea*, as was first discovered by Mr. J. V. Thompson.† From the later observations of Prof. Wyville Thomson it appears that the young Echinoderm—formed from a barrel-shaped zooid, like that of a *Holothuria*—is at first a free egg-shaped body; by the elongation of the narrow end it then becomes club-shaped, and the pointed extremity attaches itself by a disc of cement matter to some foreign body. The consolidation of the pedicle by the deposit of calcareous matter in its tissue at intervals gives it subsequently a jointed character. The rays or arms are of later growth, and do not acquire their full development till close on the period of the detachment of the body.‡

§ 5 REPRODUCTION IN THE POLYZOA.

The class of Polyzoa (Bryozoa or Bryozoaria of Foreign Zoologists) consists of minute Polype-like animals, having a great *primâ facie* resemblance to some of the Cœlenterate type, and having also, like them, a great tendency to the development of zoophytic or composite forms. Popularly they both go under the name of polypes, and even by zoologists they were confounded till a recent period. The Polyzoa, however, have a much higher organization than the

* The Holothuriæ, according to Siebold, may form an exception. He quotes Dalzell for their power of spontaneous division into two or more parts, each of which may become a complete animal, and Quatrefages for a similar multiplication by fissuration occurring in *Synapta Duvernea*. Comparative Anatomy *(Echinodermata)*, § 95, note 1.
† Edin. Philosoph. Journal, for April, 1836, p. 296.
‡ Transactions of Royal Society, Jany., 1857, and Jany., 1859.

Cœlenterate or Hydraform polypes, and one which distinctly presents the rudiments of the Molluscan type. In particular there are always distinct muscular fibres, a well-defined alimentary canal with two openings and proper walls, and traces even of a nervous system, in a single ganglion situated on one side of the oral aperture.

Their reproduction presents some interesting peculiarities, both in the protomorphic and gamomorphic stages. The early development of the germ has been followed out by Professor Allman, particularly in some of the freshwater species. By his account we have first, as the immediate result of the development of the ovum, " a ciliated sac-like embryo [germ], resembling in form and habit an infusorial animalcule. As development proceeds, we find the ciliated embryo, while still confined within the coverings of the egg, presenting in some part of its surface an opening which leads into the central cavity; and through this opening an unciliated hernia-like sac is protruded by a process of evagination. In the interior of the protrusible portion . . . . . a polypide [polype] is developed. The gemmation of the first polypide is immediately followed by that of another close beside it, so that the young polyzoon has now the appearance of a transparent closed sac, filled with fluid, the posterior part ciliated, the anterior part destitute of cilia, and partially or entirely pushed back into the posterior by a process of invagination, while the sac carries within it two young polypides, which are suspended from the inner surface of the unciliated portion."*

The prominent peculiarity here is the formation from the original germ of *two* gemmæ, which are really the embryos of the first pair of polypes.

The polypiform, which is also the orthomorphic or typical phase of development, once acquired, the tendency to gem-

* Allman's Freshwater Polyzoa (Ray Soc.), p. 41, 33, and 34.

mation becomes still more marked, and gives rise, by the continued pullulation of new zooids, to the formation of the variously branched polypidoms, which are so characteristic a feature of the class, as is indicated by the names which different naturalists have applied to it. The species are perhaps never solitary, but in a few cases they occur in pairs, the gemmation stopping short at the initial stage of the production of a double embryo.

The peculiarity referred to in the gamomorphic stage of the life history is that the structures elaborating the sexual elements are developed at so late a period, and in a manner so similar to the pullulation characteristic of the class, as to give them less the appearance of mere organs, than of distinct gemmæ or attached zooids like the polypes themselves. The fuller discussion of this point, however, must be reserved till the general relations of the organs of reproduction come under consideration.

§ 6. REPRODUCTION IN THE TUNICATA.

The Tunicata, which, along with the Polyzoa, constitute the inferior division of the Molluscan sub-kingdom, have long been known to propagate both by impregnated ova and by gemmæ. The latter are generally formed on long tubular processes emitted from the parent stock, which are equivalent to, and sometimes closely resemble, the hollow polypidoms of the Polyzoa. As in that class, too, the gemmæ frequently give rise by their cohesion to compound structures, quite distinguishable, however, by characteristic differences in the connection and disposition of the component zooids. This arrangement prevails especially in certain families, while solitary forms are more characteristic of others.

But a difference in this respect is not always to be regarded as a specific character, for it is well ascertained that in *Salpa* (and possibly in other genera) there is an alternation of two forms, from the sexual organs making their appearance only in those zooids which are derived from the original form by gemmation. It was the discovery of the connection between the two forms of *Salpæ* that first led Chamisso to introduce the expression of "Alternation of Generations;" a term since extended by Steenstrup and others to a variety of cases—some of them of very doubtful relationship. The chains which the aggregated Salpæ form by the cohesion of processes of their gelatinous integument, are modelled within the respiratory sac of the parent, on a peculiar tube, which may be compared to an introversion of the external stolon of other Tunicata. The tube, being an outgrowth from the vascular system, serves to establish a continuity between the circulation of the parent and that developed in each of the gemmæ on its exterior. The tube appears to be in a state of continuous growth from its attached extremity, for the gemmæ are most advanced at its free end. As they attain maturity, a portion of the tube breaks off from time to time with its encrustation of gemmæ, and escapes with the expiratory current, as a freely floating Salpa-chain. The aggregated *Salpæ* are bisexual, though, as the ova are much in advance of the spermatozoa, they cannot be self-impregnating, but must depend for fecundation on the entrance of the spermatic particles of other chains with the water of the respiratory current, for the development of the contained ova is well advanced before the catenated gemmæ are themselves thrown off. Each one of the *Salpæ* composing a chain matures, by the ordinary process of development, a single embryo, which eventually becomes a *Salpa* of the solitary kind, and in turn buds off in its own interior other chains of zooids like its own progenitors. From the great transparency of these animals, both forms may readily be seen at once—one within the other—and even the rudimentary beginning of a new alternation. In the living state, the different modes of connection to the parental system of the embryo and the gemmæ are at once indicated by the continuity of the circulation in the latter, and its participation in all the oscillations and irregularities of that of the adult, while the embryonic current is quite distinct, notwithstanding the vessels are curiously interlaced with those of the parent in the medium of attachment, which has very much the structure of the placenta of a Mammalian.

Changes of form also of the nature of metamorphosis occur in the development of some Tunicata. Thus the young of *Ascidia* quit the egg in the form of a *Cercaria* or microscopic tadpole *(Spinula* of Dalzell), which is afterwards transformed into that of the adult by the loss of the tail, and the evolution of the characteristic organs from the substance of the head portion.

§ 7. REPRODUCTION OF THE HIGHER MOLLUSCA.

Among the Mollusca proper, gemmation is a very exceptional phenomenon. Not only is alternation unknown at any stage of the life history, but even the implantation of the embryonic structure* on the primitive germinal mass— which is so characteristic a feature in the development of segmented animals—is here replaced by a process of transformation of the entire yolk into the substance of the embryo, and the origination of all the organs of the latter in the cells that are formed by the sub-division of the former.

The most remarkable phenomena in Molluscan embryogeny appear indeed at first sight to indicate a process of exactly the opposite kind—the fusion of numerous germinal masses into a single embryo. Koren and Danielssen have

---

* Dr. Carpenter, Principles of Compar. Physiol., 4th Ed., p. 579. In the highest class—the Cephalopoda—the embryonic structures originate from one point of the vitellus, as in most other ova.

observed that the multitude of bodies with the characters of ova, which originally fill the egg-capsules of the Pectinibranchiate Molluscs, coalesce after a time into a comparatively small number of embryonic masses, but Dr. Carpenter shows very satisfactorily that the majority of the egg-like bodies, in the nidimentary envelopes of these animals, are not true ova, but mere masses of vitelline matter, or possibly, as all undergo cleavage, though with perceptible difference, they may be unimpregnated ova, and this their last act of expiring vitality. These vitelline spheres become fused together to form a store of food-yolk, for the nutriment of the comparatively small number of embryos which are actually developed.*

In the development of Mollusca a shell is always formed, even in those species which are afterwards naked. We find also certain other organs of a provisional and temporary nature, such as a contractile caudal vesicle, and anteriorly two ciliated lobes, which serve as organs of locomotion, when the young is discharged from the egg in an immature or larva state.†

Here we can hardly avoid noticing what Dr. Burnett well calls "that most remarkable episode in the embryology of the Mollusca—the development of certain Mollusks in Holothurioidea."‡ The facts of the case rest on the authority of Prof. Müller, and are mainly these.§ In certain individuals of *Synapta digitata*, one of the Holothuridæ, there are found from one to three sac-like bodies in the general cavity of the animal, attached by their superior extremities to the head, and by their lower ends to the in-

---

* Transac. Microsc. Society, III., 17.
† Siebold, Comp. Anat., § 229.
‡ Concluding note to the Anatomy of the Cephalophora, in the translation of Siebold's Compar. Anatomy.
§ For a translation of his paper, and some judicious comments on it, see An. Nat. Hist., 2d Ser., IX., pp. 22-103.

testinal blood-vessel. At the point of connection the vessel is perforated, and the lower part of the sac so introverted on itself, that the blood can penetrate freely into the intus-suscepted portion. In the upper part of the sac are found spermatic and ovarian cysts, which discharge their contents when mature into the main sac. After fecundation from fifteen to twenty ova become invested with a common capsule. The embryos have not yet been traced to full maturity, but the course of development indicates a relationship to the Pectinibranchiata, and, as Müller thinks, to the genus *Natica*.

These facts do not seem explicable by any process of alternation on the part of the *Synapta*—even were we prepared to admit the existence of such a relation between an Echinoderm and a Mollusc—for this is not a case of gemmation but of sexual generation ; and yet it is not the true sexual generation of the *Synapta*, for this animal has the proper reproductive organs and embryogeny of its own class, and such organs have been found even in the same individual with the Molluscan brood. Parasitism affords a more plausible explanation than either "alternation" or the notion of " heterogony," which at one time suggested itself to the discoverer ; but even this view involves some startling admissions, for we must regard the sac either as a retrogressive development of a normal Mollusc, resulting in a degree of degradation—" a vermiform metamorphosis," to use Müller's own expression—unparalleled even among the Cirrhipedes or Lerneans—or else as an alternating form, budded off from a normal mollusc, for which we have as yet no precedent among the Mollusca proper. We have besides the difficulty of explaining the access of the parasite, and still more the very intimate nature of its connection with the vascular system of its host, though these are points which we can quite as little account for in some other well-known cases, unquestionably of a parasitic nature.

Mollusca, it may be remarked, do some times, though rarely, occur of parasitic habits, as, for instance, *Stylifer astericola*.

Of the gamomorphic phase of the life history of Mollusca, little need here be said. Alternation, as has been remarked, is unknown, except the presumed parasite of the *Synapta* may furnish an example. Hermaphroditism is not uncommon among the Mollusca, but self-impregnation must be rare, if it occurs at all. In some cases, indeed, it is evidently made impossible by the conformation of the parts, and in others by the sexual elements not ripening simultaneously in the same individual.

§ 8. REPRODUCTION IN THE HELMINTHA.

Among the lower vermiform tribes, the Helmintha are of especial interest for characteristic examples of the leading modifications of alternation, and for the peculiar relations of the sexes, which are here more closely associated than perhaps anywhere else in the animal kingdom. The species are frequently not only bisexual, but self-impregnating, and in some the eggs may be said to be impregnated in the very act of formation, the germinal vesicles, vitelline matter, and spermatozoa being first commingled, before they become invested with a vitelline membrane to form the ovum.*

---

* It is perhaps more correct to regard the so-called germinal vesicles with their granular investments as ova which have not acquired a proper wall, or in which, as in those of the bird, the original ovum-wall has had but a temporary existence, and to consider the matter from the so-called "vitelligenous organ," added in conjunction with the spermatozoa, as a sort of adventitious or food-yolk. See Siebold's Compar. Anat., I., § 115. Huxley in Medical Times, XIII., 133-134. Claparede (Ann. of Nat. Hist., 2d Ser., XVIII., 298 N.) Thomson in Cyclopædia of Anat. and Physiol. Supp. [120] (Ovum). In *Pelodytes hermaphroditus*, according to Schneider, in the same generative tube spermatozoids first make their appearance, and then eggs, and fecundation is effected at once. Annals Nat. Hist., 3d Ser., V., 506.

As the reproduction of the Trematoda will be afterwards noticed more in detail, in illustration of alternation in the protomorphic stage, it will suffice at present to observe that the recent researches of Van Beneden* go to show that these parasites fall into two very natural groups, distinguished alike by differences in structure, habit, and development. Some are *ectozoic*, that is, parasitic on the exterior of other animals ; they live nearly all on the gills of fishes, and attach themselves by one or even many sucking discs situated at the back part of the body. These species are viviparous, and the young, hatched from the large ova, within the body of the parent, have a development as direct as in any other animal. The second group, of which the genus *Distoma* may stand as an example, live in the interior of the body, and attach themselves by a sucker in the fore or middle part of their body. These are all oviparous. The eggs are small and very numerous, and a succession of diverse forms is very constantly interposed in the course of their development. The germ which escapes from the ovum, in the form of a ciliated animalcule, undergoes itself no farther development, but matures in its interior, and discharges a tubular sac, furnished occasionally with some rudimentary organs. Gemmæ are formed in its interior, and these, when set free by its rupture, are either converted themselves into the typical form, or give origin to others which are so. In the genus Distoma this transformation is generally effected by a gradual metamorphosis, the form first assumed being that of a *Cercaria* or microscopic tadpole, which, losing its tail, is eventually transformed into the perfect *Distoma*, during an encysting process which the parasite undergoes, immediately on penetrating into the tissues of living animals. Sexual organs are

* See Translation of Van der Hœven's Abstract, in Annals of Nat. Hist., 3d Ser., III., p. 344.

not developed till some time after the distoma form has been fully acquired.

The series of changes is still more complicated in the case of many of the Cestoid worms,* and the study of their development has led to the discovery that the Cystic worms, once supposed to be a distinct order of Entozoa, are merely provisional forms, belonging to the early progress of species of the Cestoid division. The egg of a Tape-worm, or other Cestoid Entozoon, gives origin to a minute contractile vesicle, armed with six hooks, by which it is enabled to bore its way into the tissues of animals. When it is once established in suitable quarters, the primary cyst buds off what must be regarded as the typical form of the order—the *Tænia-head* —characterized by its four suckers and apical circlet of hooked teeth. By a continuation of the budding process, the Tænia head may be raised on the extremity of a hollow jointed pedicle forming a flexible neck, but no development of proper reproductive organs takes place, so long as the parasite continues in the same locality. On its being transferred, however, to the alimentary canal of a warm-blooded animal, the original caudal vesicle is cast off, and a series of new segments are budded off from the hinder part of the Tænia head. As the gemmation is continuous, and the segments first generated remain adherent to those of subsequent formation, by whose outgrowth they are pushed off from their point of origin, a long jointed appendage is developed, extending backwards from the head. This new formation constitutes the "body" of the Tape-worm, and it is in its segments that the reproductive organs eventually make their appearance. But the ultimate segments are not all *directly* derived from the head, as a process of transverse sub-division comes in to supplement the

* According to Van Beneden, direct development, without alternation, occurs only in *Caryophylleus*, a genus inhabiting the intestinal canal of certain fishes. Annals Nat. Hist., 3d Ser., III., 346.

original gemmation. The joint next the head is soon divided by a transverse fissure into two, each of which repeats the process as soon as it is somewhat grown. Whilst the joints multiply in this way they increase in size in the same proportion, and so, of course, remove the joints from the head.* But at a certain distance from the head, the whole nutritive power is applied to the development of the organs of generation, and this mode of sub-division ceases, though the budding off of new segments from the head continues. The segments are at first very minute, but as their growth now becomes rapid, they soon come greatly to exceed the size of the head from which they were originally derived. It is this enlargement of the segments, when they have been thrust to some distance from the head, that gives rise to the peculiar form of the neck of the tapeworm—attenuated to a mere filament where it joins the head, but thickening behind, as it passes insensibly into the body. Organs of both sexes occur in the same segment, but they do not begin to make their appearance till the joints have acquired considerable dimensions, and are always found more mature as we pass towards the hindmost—that is, the first-formed segments of the body. When impregnation has taken place, and the ova are ready for evacuation, the segments break off, and are discharged with the fœces, still retaining a certain degree of contractile vitality, which aids in the dispersion of the contained ova.

In no species probably of either kingdom are all the three stages before distinguished as protomorphic, orthomorphic, and gamomorphic, more clearly marked than in the Cestoid Entozoa. To the first belongs the contractile six-hooked vesicle discharged from the egg, to the second the Tænia-head, and to the third the jointed body of the Tapeworm. Of these three successive phases, each of the later is derived

* Eschricht on the Generation of Intestinal Worms. Edinburgh Philosophical Journal (Oct., 1841), p. 340.

F

from the preceding by a distinct gemmation; yet, from the gemmæ remaining adherent to the stock, the process has so much the character of a continuous growth, that only *two* alternating forms are generally recognised—the Cystic and the Cestoid—and even these owe their distinction as much to the disappearance of the caudal vesicle of the former as to any *formative* act. Hence the Tapeworm is often described as a tænioid entozoon which has lost its original cystic appendage, and developed a long cestoid one in its stead. But in no proper sense can the caudal cyst be termed an appendage of the Tænia-head—the head is rather an appendage of the cyst, as developed from it. The Tænia-head is indeed the one common feature that unites the two forms, but it stands in a very different relation to the one and to the other; it is the *offspring* of the cyst— the *parent* of the cestoid body. A sort of inversion occurs in the direction of the gemmation during the course of the genetic cycle. It is forwards till the head is formed, when it is reversed, and goes on subsequently in a backward direction, so that, had the cyst not been thrown off in the meantime, we should expect to find the whole length of the jointed body interposed between it and the head. Something of this kind would seem actually to occur in certain species,* and such is the normal arrangement in the class of Annelida.†

In the third order of these Entozoa, the Nematoid worms, alternation is certainly not the general rule, though there are grounds for admitting its occurrence in some species. Thus of late reasons of weight have been brought forward to induce us to look for the progenitor of the formidable Guinea worm in a microscopical inhabitant of tropical pools —the Tankworm—of such tenuity as to be capable of

---

* Especially in Tetrarhynchus, according to Van Beneden.
† Carter, Annals of Nat. Hist., 3d Ser., I., 299; IV., 33-99.

effecting an entrance into the sudoriparous ducts of the skin. Other *Filariæ* which do not develope sexual organs may probably have a like origin. In the same way Kuchenmeister and other writers are inclined to identify the *Trichina spiralis* with the *Trichocephalus dispar*; but in regard to this there is greater dubiety, as there is a certain amount of evidence in favour of the *Trichina* developing sexual organs of its own.*

§ 9. REPRODUCTION IN THE ANNELIDA.

Though egg-like bodies with more than one embryo have been met with in some Annelida, the probability is that they are only nidimentary envelopes, such as are common among the Gasteropoda, for we have no evidence of any fissiparous or gemmiparous multiplication in this class, in the protomorphic stage of development.† With the first

* For a list of references on the development of *Trichina*, see Burnett's note in his Transl. of Siebold's Compar. Anat., p. 31.

The most recent researches are those of Leuckart, communicated in the present year to the Royal Academy of Sciences of Gottingen. The results are as follows:—

The *Trichina* is the young state of a small Nematoid worm hitherto unknown, but occurring in large numbers in the intestines of many Mammalia and Birds. It is introduced by the ingestion and digestion of the flesh of the prey affected with the parasite, and very soon when set free acquires full sexual maturity. The young are developed in about six days, and immediately begin to make their way to the muscular tissues, by penetrating through the wall of the intestine, and the peritoneal covering of the abdominal cavity. They make their way finally into the interior of the ultimate muscular fibres, and there attain within fourteen days the size and organization of the *Trichina spiralis*, as commonly met with. The penetration of these embryos in large quantity sometimes gives rise to dangerous peritonitis. Annals of Nat. Hist., 3d Ser., V., 504.

† In some Nemertini, according to Desor and Schulze, the early development somewhat resembles that of the Trematode Entozoa, for the first formation from the egg is a ciliated infusorian, with a mouth-like cleft on one side. Within this is generated an active vermiform body, which eventually escapes from the matrix, carrying away the cleft mentioned,

appearance, however, of the embryonic structures, gemmation comes into play, manifesting itself primarily and most extensively in the multiplication of the segments of the body. The species with external branchia quit the egg as short ciliated larva-like Infusoria, and only acquire the vermiform character by the successive gemmation of segments from behind, much as the joints forming the "body" of the Tapeworm are developed from the back of the Tænia-head. Only there is this difference, that, with the exception of the caudal appendage, the segments are budded off from each other, not all from the head, so that the penultimate is always the most recent, while in the Cestoids the posterior segments are the oldest.

In both cases it is only in these derivative segments that generative organs are developed, and the parallel is carried still farther, when, as in *Syllis*, they are thrown off as sexual zooids, the main difference being that in the Annelida the offsets have more the character of distinct animals. The joints are not only thrown off in sets, like the catenated zooids of the *Salpæ*, but such an organic unity is established among the connected segments as to give each set quite the character of a complete animal. In these cases one or more of the later segments become secondary foci of a budding process, which proceeds exactly in the same way as in the young annelidan first formed from the ovum ; that is, the ring which is to develope the new zooid sub-divides into two parts, which acquire the organization respectively of head and tail segments, but it presents at first no farther division, the other joints being gradually formed afterwards, between the cephalic and caudal extremities, and always in succession from the posterior of the segments previously produced. Phenomena of this kind were first observed by

as its own mouth. From this body the annelidan is probably developed, after the plan which prevails generally in the class. Huxley in Medical Times, XIII., 281.

Otto Müller in *Nais proboscidea*, and by Gruithuisen in a species of *Nereis*. More recently M. Quatrefages has made similar observations on the *Syllis prolifera*. This species forms ordinarily but a single zooid at a time, but Mr. Milne Edwards in another annelidan—*Myrianida fasciata*—has seen as many as six in process of formation at the same time from the terminal segments of the parent. The first formed and most complete was situated furthest back, and each newer zooid presented a less developed structure than the preceding one. The anterior or youngest had only ten rings, the second had fourteen, the third sixteen, the fourth eighteen, the fifth twenty-three, and the last or caudal one thirty rings. In some of the species named, the fissiparous multiplication has very markedly the character of a process ancillary to sexual reproduction, for the zooids when thrown off speedily develope the appropriate organs, but retain life only long enough to impregnate or mature the resulting broods of ova.* In other cases, perhaps, multiplication may go on indefinitely, by the detached zooids repeating the same fissiparous process—the formation of sexual organs appearing to depend in some degree on external circumstances. Multiplication may also be effected by artificial division, in some cases in which the fissiparous process does not seem to come into play as a spontaneous mode of increase.†

In the sexual relations of the Annelida there is considerable variety, hemaphroditism being a common character among the Terricolæ and Suctoria, while the sexes are generally separate in the other orders. There is no evidence of self-impregnation in any.

---

* Quatrefages Ann. Des. Sci. Nat. (1844) T. I., p. 22. Edwards, Op. Cit. (1845), T. III., p. 170. Dr. A. Thomson, Cyclop. Anat. and Physiol., Art. Ovum., p. 32-33.

† Siebold, Compar. Anat. I., § 163.

§ 10. ARTICULATA.

Among the higher Articulata, it is only in a few aberrant species that we have any obvious manifestation of gemmation; consequently, though in these exceptional cases, we meet with some well marked examples of "alternation," such phenomena are not in accordance with the general course of development in this type of organization.

It may be necessary in some degree to qualify this statement, as applicable to the Myriapoda, for as these approach the Annelida in configuration, so they do also in multiplying their segments up to a certain point, by a process of gemmation from the penultimate segments, subsequent to the emission of the embryo from the egg, though no instances are known of spontaneous fission of the body.*

A striking feature in the reproductive process among Articulata is what is called *metamorphosis*. In Insects particularly, and also in many Crustacea, the embryonic form first assumed often differs very widely from that of the adult, in the conformation of its parts both external and internal, and this not merely in its organization being rudimentary, but in a total dissimilarity of particular organs or members, and even in the existence of structures which afterwards waste or entirely disappear by absorption, or are thrown off from the body. Metamorphosis, indeed, is not confined to this division of the Animal Kingdom; allusion has before been made to its occurrence in some of the other groups, and, as a general rule, it may be said to be more marked in proportion as the typical organization becomes more elaborate. It is nowhere, however, more observable

---

* According to Fabre, the larva of *Scutigera* has at first only the seven anterior segments, containing the organs of sense and the digestive apparatus; the posterior segments—those behind the large plates or elytra—are of later formation. It is in this region, as in the case of the Annelida, that the reproductive organs are situated. Annals of Nat. Hist., 2d Ser., XIX., 165.

than in the case of Insects, owing partly to the real magnitude of the changes which take place, and partly also to the prominence with which they are brought under the eye of the observer. The embryo even of Mammalia undergoes in the course of its development a series of remarkable changes, very parallel to the metamorphosis of an insect; the conformation of its vascular system, for instance, is changed more than once, and certain provisional organs, such as the wolffian bodies, the umbilical vesicle, and the allantois, appear, which are afterwards absorbed or cast off. But these transformations have only been traced by laborious anatomical investigations, as they are all gone through in the interior of the uterine cavity—in those hidden recesses where the new being is "made secretly, and fashioned beneath in the earth," as it were—while in the Insects and some others of the lower animals, the embryo being extruded, not only from the body of the parent, but also from its own egg-coverings, while still in an unformed or larva state, the whole subsequent course of metamorphosis takes place in the outer world, and is in great measure presented naked to our view. Larvæ are not unknown in other forms of animal life—even among Vertebrata they occur in the case of the Batrachia and Marsupialia, and also in some Fishes*—but in Insects they are not the exception but the rule—the guise under which the young are normally brought forth.

Entomologists distinguish three stages in the development of Insects, by the terms *larva*, *pupa*, and *imago*. But these expressions do not always indicate homologous conditions, for the point of development at which the young quits the egg varies considerably in different orders of the class. "In all cases the germinal mass, while still within the

* The metamorphosis is very marked in the Lamprey. Müller's Archives Anat. Physiol. (1856), p. 323.

egg, is first converted into a footless worm, resembling the higher Entozoa, or the inferior Annelida, in its general organization, but possessing the number of segments—thirteen—which is typical of the class of Insects."* It is at this stage of its progress that the embryo of the Diptera, Hymenoptera, and of some of the Coleoptera or Beetle-tribe, comes forth as a larva, such as is commonly known as a "Maggot." In Butterflies, again, and in the greater number of Beetles, the organization of the larva is rather more advanced, as it is furnished with the rudiments of the thoracic legs, besides certain provisional organs of the same nature on the posterior segments of the body—the abdominal tubercles or pro-legs. Such larvæ are termed "Caterpillars." In the Orthoptera, Hemiptera, and a portion of the Coleoptera, there cannot be said to be any proper larvæ, as the young do not emerge from the egg till they have attained the conformation of the adult in almost all respects, except the evolution of the wings. The true or vermiform larva feeds voraciously, and continues for some time without any other obvious change than a rapid increase in size, attended generally with several castings of its skin. As the period of its transformation approaches, it passes into the *pupa* condition—a state of inactivity, during which it undergoes its change into the *imago*, or perfect insect, with six jointed legs, four wings, and other peculiarities of that type of organization. "The pupa is enclosed in the last skin exuviated by the larva, which, instead of being thrown off, dries up, and remains to encase the proper skin of the pupa that is formed beneath it; and, in addition to this, it is frequently protected by a silken 'cocoon,' the construction of which was the last act of larval life."† To this association of external inactivity, with great internal constructive energy—that is, to the

* Carpenter, Compar. Physiol., 4th Ed., p. 599.
† Carpenter, Op. Cit., p. 601.

pupa state in a physiological sense—that which corresponds in the case of the Orthoptera and Hemiptera, is the later period of development within the egg. In other words, these embryos pass through stages, corresponding both to larva and pupa within the egg, or at least are only discharged while still undergoing some of the concluding changes of the latter state. Mr. Andrew Murray brings forward evidence to show that in some cases even a sort of cocoon is formed within the egg, and that the young insect, after certain changes, emerges from it, before it is finally discharged from the egg.*

In the lower orders of Insects, the development is more or less arrested; in the Anoplura, and Thysanura, which come out of the egg as active pupæ, there is no farther metamorphosis, and in the Homoptera, it is so far imperfect that but two wings are formed; in the flea tribe also, though the larva and pupa states are regularly gone through, the development of the *imago* form is imperfect, the abdomen not being marked off from the thorax, and the wings being represented only by rudimentary appendages attached to the second and third segments of the body; in Diptera the posterior wings are in like manner rudimentary. In other cases it is only one of the sexes, or it may be only certain individuals of the sex, whose development is so arrested. Thus the females of *Coccus* and *Lampyris* never develope wings, and in the ant tribe neither these organs nor those of reproduction are evolved in the majority of the sex, full development being attained by a few individuals only, which breed for the whole community. Among the *Termites*, or white ants, the "soldiers" appear to be pupæ arrested in their development, while the "workers" have the characters of permanent larvæ. The working females of the bee, on the other hand, acquire the external charac-

* Ed. Phil. Journal. 1858.

ters of the perfect insect, only their sexual development being arrested. In this they contrast with those cases in which the males are the subjects of arrest, as among the Cirrhipedes and Lerneans; for then it is the general organization that is defective, the sexual parts being developed out of all proportion to the rest of the body.

The whole development, however, of these last-mentioned orders is exceedingly anomalous. The first change, induced on the germ-mass of the ovum, assimilates it to the general character of Entomostraceous Crustaceans, but a subsequent transformation, of what is called a retrograde character, results in the loss of vision and of the power of locomotion, and so masks the articulate character that the forms ultimately assumed were long referred to wholly different types of organization. There is, however, another possible view of such cases, which will be considered afterwards.

Throughout the articulate series, the sexes are normally separate, except in some aberrant forms of the Crustacean class, and perhaps also some of the lower Arachnida.

The cases of well-marked alternation, such as occurs among the *Aphides*, will be taken into consideration when that subject comes to be discussed.

§ 11. VERTEBRATA.

The remarks just made in regard to the abeyance of gemmation, the absence of phenomena of alternation, and the constancy of the separation of the sexes in Articulata, all apply with still greater force to the Vertebrate division of the Animal Kingdom; but no farther remarks seem to be called for here on the reproductive process in this group, as the relation to it of various modes of propagation occurring among the lower organisms has been throughout a primary object of attention in this summary.

§ 12. An observation that suggests itself in comparing the reproductive process in the several groups of Animals, as represented in the preceding summary, with the modifications of the same functions of Vegetables, is the singular difference of the general result in the two kingdoms, arising from the same general law of the interpolation of gemmation at certain points in the genetic cycle. Its effect in the case of plants is to establish a wide apparent diversity between the reproductive process of the higher and the lower cryptogamic plants; while in animals it is rather to bring forward, in the case of particular families or species, a series of forms, which, however apparently dissimilar, are really only successive phases of one individual, or detached offshoots or members of the same original stock.

Very much of the general character of the reproductive process in any family depends on the mutual relations of certain epochal acts, as they may be called—such as access of the spermatic element (Insemination=I); discharge of the ovum or its contents (Extrusion=E); escape of the young from the egg coverings (Birth=B); and the full acquisition of the typical characters (Development=D). The following tabular statement, therefore, in which these acts are denoted by the letters above indicated, may be of interest, as showing some of the more usual variations, in their order of succession, in the different tribes of animals. When two letters follow each other immediately, it denotes that the acts are simultaneous; the interposition of a greater or less number of points indicates a corresponding interval of time in the succession :—

Mammalia............ I .... DEB
Marsupialia......... I...EB .... D
Birds................ I.E .... DB
Batrachia............ EI...B .... D
Fishes............... E.I...B .... D
Insects.............. I.E...B .... D

There would be considerable difficulty in extending such a table to the Invertebrata generally, from the want of uniformity in many of the orders, in the points referred to.

IV.

THE NATURE AND VARIETIES OF ALTERNATION OF GENERATIONS.

§ 1. THE two modes of propagation—by gemmæ capable of spontaneous evolution, and by germs dependent on impregnation—as has been already observed, are frequently associated with no less remarkable diversities in the immediate result of the development, leading in cases of periodic recurrence or alternation of the former, to a corresponding mutation or alternation of dissimilar forms in the same species. It is only, however, quite recently that this has been admitted generally by zoologists, who were not unnaturally indisposed to it, by observing the constant succession of like to like in the higher animals. But since the time that Chamisso called the attention of naturalists to the recurrence of two forms in *Salpa*, as a case of "Alternation of Generations," analogous phenomena have been abundantly brought forward in other tribes of organised beings. Steenstrup was the first to group together these cases, applying to them the same term as was used by the former naturalist, for which some later writers would substitute that of *Metagenesis*, proposed originally by Professor Owen.

In all these cases we may admit so much as this in common—that an act of digenesis recurs with greater regularity in the interval of the acts of monogenesis ; and that the products of the former differ more or less in their conformation from the organisms budded off in the latter.

Hence, as both forms must be taken into account to complete our idea of the perfection of the species, it has been proposed to term them zooids in the case of animals, and phytoids in that of plants, as indicating that any one of them is not so much a complete animal or plant in itself, as a fragment or fractional part of one—the whole series, considered as specific unit, rather than any one among the successive links of which it is made up, answering to our idea of individual completeness, as this is drawn from the higher animals, in which like seems always to produce like.

In confirmation of such a view, it is noticed that in not a few cases these fractional phytoids and zooids really remain in organic union for life—making up an arborescent form—like what we call a polypidom in animals, which is readily recognized as being in its entireness the individual representative of the species.

In this view *zoological* individuality becomes a very different thing from *metaphysical* individuality,* for organs, which in one species are integrally united into a whole which is indivisible without mutilation, are found in others more or less broken up into separate and independent structures. Hence if we would avoid the solecism of speaking of the *individual*, as in some cases being *divided* into a number of parts, having all a general character of completeness, and use the word, as propriety seems to require, in its logical sense of *that which is individualized*, we must be prepared to admit as a consequence, that the individuals of one species are not always homologous with those of another.†

But though we may allow so much in common in these cases of "alternation," as is involved in the occurrence in all of a periodic diversity of derived forms, there are yet—as was pointed out in the introductory chapter—great variations among them, as far as the relations are

---

* Huxley in Philos. Transactions for 1851.   Paper on *Salpæ*, p. 579.
† Mr. Lubbock in Philosophical Transactions.   Jany. 29, 1857.

concerned in which the budding process stands to the sexual act, and to the full development of the specific type—relations depending on the period of the life-history of the species, at which the act of gemmation is interpolated in the genetic cycle. The contrast lies especially between the cases in which the alternation of form is due to zooids being budded off in the *Protomorphic* stage of the life-history—that is, during the early progress of germinal development—and those in which it arises from the detachment of gemmæ in the fully developed or typical phase, as a preliminary step to the evolution of reproductive organs —the latter zooids belonging to the *Gamomorphic* stage, or that of sexual maturation. The two classes—as has been already observed—differ widely in their structure and relations. In the one case they are the primary products of impregnation, precursors of the perfect form, and without sexual characters—in the other derivative, and with distinct sex. Zooids of both kinds, indeed, may have certain organs superadded, varying in their nature and completeness with the circumstances of their life as independent beings. In those of the protomorphic stage, the adventitious organization probably does not go beyond the development, externally, of cilia, or of a contractile integument for locomotion, and internally, of a rudimentary digestive apparatus; but in many gamomorphic zooids, both the locomotive and alimentary systems may be rather highly organized, and the whole structure occasionally larger and more complex and elaborate than that of the parent stock. On the other hand, such is the structural degradation of some zooids of both kinds, that they might readily pass for mere proliferous cysts or egg-sacs. This variability in the kind and extent of organization proves of itself its adventitious nature, and shows it to be of no value as a distinctive feature. The real points of distinction are those before referred to—their position in the genetic cycle, and their gemmiparous or sexual character in consequence. They both, however, have this in common, that the great end of their existence is the multiplication of the race—an end to which the nutritive and animal functions are always subordinated.

§ 2. The distinctions now referred to among the cases of alternation will appear more clearly by noticing the phenomena which are actually met with in such tribes of animals as afford well marked examples of their occurrence. The case of the Trematode Entozoa furnishes a fair illustration of the variety in which gemmation occurs, as a feature in the progress of the development of the most typical form of the species. The reproduction of these Entozoa, as observed in the well-known parasite *Distoma*, has had the attention of naturalists strongly drawn to it, from being the case on which Steenstrup chiefly enlarged in his original work on "Alternation of Generations." The species of *Distoma* are very numerous; they are bisexual, as is the case with the Trematoda generally, and probably also self-impregnating. The young, on escaping from the egg—which does not commonly happen till the ejection of the latter from the system of the animal in which the parent was a parasite—appear in the guise of infusorial animalcules, which swim about freely by the play of the cilia, covering their exterior. Some of them have been traced into the interior of water snails. This infusorial form never advances to any higher development; it is in fact merely the matrix of another animalcule of a very different character, resembling a *Gregarina*, in its uniformly granular structure and smooth contractile integument, and the function of the original ciliated envelope is mainly, as it would seem, to introduce the *Gregarina* into the body of the snail. This gregariniform parasite itself is a mere cyst or capsule, whose gelatinous walls have generally a certain degree of contractility, but which never acquire more than the most rudimentary internal organs, though there is a certain gradation observable in this respect. In some there is a distinct head, with a mouth opening into a sort of blind gullet, representing perhaps the commencement of an alimentary canal, and posteriorly the tail is marked off from the cylindrical body by a pair of processes like rudimentary limbs; others again are merely elongated sacs, not always endowed even with contractility. The former Filippi terms *Rediæ*, the latter *Sporocysts*.*

In the farther course of development, the redia or sporocyst forms in its interior a number of growing points, which assume the form of *Cercariæ*, or microscopic tadpoles. The parent cyst in time gives way, under the pressure caused by the growth of its progeny, which are thus set at liberty, and boring their way out of the snail by hooks on their heads, they swim about for a time freely in the water. Their relations and farther progress are thus graphically described by Prof. Owen. "No sexual organs exist in these *Cercariæ*, any more than in their animated 'coat,' the *Gregarina*, or in their ciliated 'great coat,' the Infusorial embryo. After the larval Cercariæ have passed some time in the water, first creeping and then swimming about with great restlessness, they either enter directly the body of the waterfowl,† or bore their way into some aquatic insect, or they may fail in both these instinctive efforts, and remain in the water. In any case they undergo a metamorphosis. The Cercaria gathers itself up into a ball, and exudes a mucous secretion from its surface, which soon hardens; and since the worm, inside this mucous mass, turns round without stopping, it

---

* Siebold's Compar. Anatomy, § 118, n. 7. Huxley, in Medical Times, XIII., 133-134. Busk's Translation of Steenstrup on Alternation of Generations, (Ray Soc. Pub.), p. 63.

† The *Monostomum mutabile*, a Trematode, whose development is quite parallel to that of *Distoma*, has for its habitat the eyelids of ducks and other waterfowl.

invests itself with a kind of egg-shell; during this process the tail is cast off.*

"Should this process take place within the body of an insect, the encysted Cercaria might be introduced into the body of an insectivorous bird or beast. In the act of digestion by the engulpher, the body of the insect is destroyed, together with the capsule of the cercarian pupa; but this by virtue of its vitality remains unharmed, and is thus transplanted into a new sphere, and fitted for its farther change into a sexual entozoon of the Trematode or 'fluke-worm' order.

"Then again commences the strange and complex genetic cycle, from the Harveian point—the impregnated ovum.

"Three different species of animal may contribute—two are essential—to the successful progress of the ordinary and Parthogenetic processes of propagation, manifested by the three distinct forms of Infusory, Gregarina, and Cercaria, intervening between the egg and the perfect parasite fluke-worm."†

If the hypercriticism be excusable of objecting to the accuracy of the figurative language, in which this author gives so lively a representation of these strange mutations, it may be remarked that the terms "coat" and "greatcoat" are out of point here in two particulars.

1. That such articles of dress never envelope more than one wearer at a time, whereas the Gregarina-form at least encloses a whole brood of Cercariæ; and

2. That these garments are worn simultaneously, whereas the Gregarina does not develope the Cercariæ till it has escaped from its infusorian cyst. Indeed in some cases it

---

* Sometimes this appendage appears to be torn off at the first entrance of the parasite, in the act of boring through the integument. It would appear, however, that some Distomata never pass through the Cercaria stage at all. Huxley, in Medical Times, III., 133-134. Siebold's Compar. Anatomy, 118 [note].

† Address to Brit. Assoc., 1858, 23-24.

would seem that the Cercariæ are not formed in the first Gregarina-zooid at all, but in those of an intermediate brood, derived from it, and of the same general character.*

Till the recent researches of Professor Van Beneden of Louvain, the whole series of changes had never been traced out in the same species, or by one single observer, but, as the author just quoted remarks in another place,† this testimony of different good and independent witnesses, at different periods, to different stages of the successive generations, which, when compared with one another, are seen to link together and complete the metagenetic cycle, was perhaps the best foundation for scientific confidence in the truly marvellous result.

"Siebold observes the development and birth of the ciliated monadiform embryo from the egg of the oviparous Trematode Entozoon, and the escape of the gregariniform non-ciliated worm from the ciliated one.

"Bojanus and V. Baer trace the development of the cercariform individuals from the 'King's yellow worm,' which in its form and simple structure corresponds with the vermiform offspring of the ciliated embryo.

"Nitsch traces the Cercariform progeny of that worm to their pupa state.

"Steenstrup, confirming the origin of the Cercariæ from the multiparous 'King's yellow worm,' completes the observation of their metamorphosis through their pupa state, into the Trematode Entozoon."

In the development of the Trematoda, as exemplified in the case of a *Distoma*, there are two points of especial interest in their bearing on the general question of Alternation:

1. The imperfection and variability in the amount of organization acquired by the zooids, which originate directly from the infusorian form, as compared with that of the

---

* Steenstrup on "Alternation," pp. 69-93.
† Parthenogenesis, p. 17.

Distomata, derived from them in turn. The former never exhibit clearer indications of animality than a power of peristaltic motion, and some rudimentary organs, external and internal; and they range from this down to simple rigid tubes—mere bud-sacs, quite destitute of organization and all power of motion. The latter, again, possess as complex a structure as is found anywhere in the class of animals to which they belong—and this not as an occasional development, but uniformly in all cases, and quite irrespective of the organization of the precursory form from which they have been derived. The derivative zooid, indeed, is not only much more highly organized than the primary, but it is truly the *typical* form of the species, as appears by its correspondence with the fully developed condition of those Trematoda in which "alternation" does not prevail, and consequently but one form occurs in the same species.

2. The second point is that not only are the precursory zooids destitute of all sexual characters, but that no such organs are developed even in the derivative forms, till sometime after they have acquired the typical conformation in all other particulars. The appearance of these organs, in fact, constitutes a new epoch in the life-history of the typical form, quite as observable as its derivation from the preceding one, and admits of our dividing the whole succession of changes into the three stages to which the terms Protomorphic, Orthomorphic, and Gamomorphic, were applied in the Introductory Chapter.

From the consideration of these points, we may assign the following as the characters—positive and negative—of that modification of "alternation," which prevails among the Trematoda :—

1. That the "alternation" consists in the derivation of sexual and oviparous zooids, from the last generation of a series of others which are sexless and gemmiparous.

2. That the later set of zooids, which eventually acquire

sexual organs, represent the typical or orthomorphic form of the species.

3. That the process of gemmation in this case represents a particular step in the course of embryonic development, marking the transition from the early germinal or protomorphic, to the first beginning of the orthomorphic or typical stage; and that it has no direct bearing on the evolution of the sexual organs—the process which distinguishes the gamomorphic stage.

The first of these characters belongs to all forms of " alternation ;" the latter two are distinctive of that now described, which may, therefore, be termed " Alternation from gemmation in the Protomorphic stage," or, more shortly, " Protomorphic Alternation." Its distinctness will appear at once, on comparing it with other forms, particularly with that which prevails among the lower or polypiferous section of the Cœlenterate division of the Animal Kingdom (Hydrozoa of Huxley).

§ 3. In more than one family of Polypes, the true ova are frequently produced by zooids, varying much in size, but having the general structure of the Medusæ, or common Jelly-fishes of our seas. From the ovum a cup-shaped embryo proceeds, which is transformed into a polype, and from this, as from the original shoot of a plant, there is generally developed, by a series of successive sprouts, a polypidom supporting a whole colony of polypes like the original—variously disposed according to the species, but all organically connected together. At certain seasons, and under favourable circumstances, there are developed either from the polypes or from the polypidom, peculiar bud-like processes, which sometimes acquire the characters of regularly organized Medusoids, but at others remain in the rudimentary condition of mere spermatic and ovarian sacs. It is from these that the ova are derived, whose course of development has just been noticed.

Now, while this case agrees with that of the Trematoda, in the first of the three characters before given—that of a derivation of sexual and oviparous zooids (the medusoids) from others which are sexless and gemmiparous (the polypes, or their common stem, or its capsular developments)—I think it may be clearly shown to differ in the other two ; that is to say, in the zooids which represent the typical form being here the earlier or gemmiparous set, not the later or oviparous ; and in the process of gemmation being ancillary to the act of sexual reproduction, instead of representing a step in the development of the typical form—belonging to the Gamomorphic, instead of to the Protomorphic stage.

Certainly it is the polypiferous phase which in the great majority of the species impresses us as the representative of the typical form, for in the Coryniform and Sertularian zoophytes, as well as in the Calycophoridæ and Physophoridæ, the individual polypes, though small, are sometimes not so minute as the resulting medusoids ; while the zoophytic growths, which are developed by their repeated pullulation, are clearly the most conspicuous phase in the life-history of the species, and, in the two orders last mentioned, have often such individuality of character about them, that they suggest the idea not so much of a colony of aggregated zooids, as of a body composed of various organs, like that of one of the higher animals.*

And as in these cases the polype phase is the most conspicuous, so is it always the most permanent condition, for, after throwing off a swarm of medusoids, it goes on in as vigorous a state as before, and at intervals, varying with the species or with accidental circumstances, may give origin to

---

* This is especially the case in such forms as *Physalia*, *Velella*, and *Porpita*.

many successive crops of gemmæ* ; whereas the rudimentary forms of the Trematoda seem to pass away as soon as they have thrown off all their freight of gemmæ, whose production is the great object of their existence. The contrast is equally marked in the derivative zooids, though in the opposite way. In the Trematoda, where they are represented by the distoma-form, it is these that have the greater length of life ; for a considerable time generally elapses before they go through their metamorphosis and acquire sexual organs—while the medusoids of the zoophytes, which are developed for the very purpose of reproduction, have a correspondingly brief duration, their term of life coming to a close as soon as the spermatic and germinal products are matured and discharged.

There is besides a fragmentary and subsidiary character attaching to these medusoids in many cases, which of itself suggests the idea of their being merely detached organs for the formation and fecundation of ova. It is true that there are great differences among them in this respect—greater even than among the precursory zooids of the Trematoda. The variability is such that while some of the medusoids have an organization apparently much more complex than that of the polype-stock, in other cases the sexual zooids are reduced to mere sperm and germ-sacs, of the most rudimentary description, without any trace of medusoid structure, though the correspondence of the two extremes is established by a continuous series of intermediate forms.

There is an equal variability in the isolation or attachment of the sexual structures. In general these are thrown off as free zooids, but in many species they are never detached, forming fixed appendages of the parent stock, and gradually merging into the character of mere organs of reproduction. So completely is this the case in the common *Hydra*, that

* Carpenter's Compar. Physiology, 4th Ed., p. 559.

the species presents but a single form, very slightly modified even at the breeding period. The single form of the *Hydra* may, of course, be as fairly taken to indicate the type of the allied polypes, as that of the viviparous Trematoda to determine the type of the order of worms to which they belong. So long, therefore, as we confine our attention to the orders of Hydrozoa above mentioned, there seems no difficulty in determining that here the "alternation" is due to gemmation in the gamomorphic stage, or in the course of the evolution of the sexual organs—that the oviparous zooid is not here the typical form, but a subsidiary offset from it, for the development of those organs of reproduction which never make their appearance in the former—and that the gemmiparous zooid is not here the representative of the gemmiparous or protomorphic zooid of the Trematoda, but is itself the typical form, which in polypes is produced by the *transformation* rather than the *gemmation* of the primary infusorial product of the ovum ; or, as we recognise in both cases a principal or typical and a subsidiary form, the difference might otherwise be expressed by saying, that in the Trematoda the subsidiary zooid is gemmiparous and preliminary to the typical, while in the Polypifera it is oviparous and supplementary.

But there is one order of Hydrozoa—the Lucernariadæ—some of the reproductive phenomena of which seem at first sight much opposed to this view. Certain Lucernarian polypes, particularly that termed *Scyphistoma* by Sars, *(Hydra Tuba* of Dalzell) give off gemmæ, which are eventually developed into Hood-eyed Medusæ—the common Jelly-fishes of our seas—animals of colossal dimensions, compared with the original polype stock, and, apparently at least, of much higher organization.* Here it seems natural

---

* On the Lucernarian affinities of the polype phase of the large Medusæ, see Huxley in Med. Times, XII., 506, and Oceanic Hydrozoa, p. 21. See also Dr. J. Reid's Physiolog. and Patholog. Researches, p. 445.

to consider the polype form as a rudimentary phase, comparable to that occurring among the Trematoda, and to regard as typical the Medusa form, which is an animal of much more conspicuous appearance, of apparently higher organization, endued with the power of locomotion, and having true sexual organs in distinct individuals. Such, indeed, is the view taken by Steenstrup and several other authors, even by those who occasionally speak of the medusoid progeny of the Hydraform polypes as mere detached generative organs. Thus Prof. Owen observes—" The nutritive gemmiparous polypiform individuals in all the compound Radiaries would seem, rather than the oviparous medusiform ones, to manifest the typical form of the species. . . . Superadd, however, distinct nutritive and circulating organs to the free-moving ovigerous individual from the rooted polype, and prolong its existence, and it would then cease to have the ancillary character of a nurse to the ova of the fixed individuals, and would assume that of the perfected form of the species ; and such in fact is the case with the larger gelatinous Radiaries called Medusæ." *

Now, in so far as perfection means the possession of a more conspicuous organization, it is not of course to be denied that the Medusa is in advance of the polype ; but as regards the selection of the phase to be taken as the typical form of the species, I do not see how we can avoid these conclusions—

1. That the Medusoids derived from the compound polypes (Sertularida and others)—formerly ranked as a distinct order under the name of Gymnophthalmatous (or bareeyed) Medusæ—are really homologues of the parts of reproduction, inasmuch as they pass by a continuous gradation into generative organs of the simplest kind.

* Parthenogenesis, p. 12.

2. That in so natural a group, the relative position assumed for the puny bare-eyed Medusæ must hold also for their portly brethren of the hood-eyed kind.

A few details may here be given of the declension referred to, from free Medusæ to germ-sacs of so simple a kind, that they might fairly be termed mere tunicated ova. The transition is so gradual, that Prof. Huxley considers it impossible to draw any definite line of distinction between true Medusoids and mere proliferous cysts, or sporosacs, as they are termed by Prof. Allman.* The distinction, however, is a convenient one, if not strictly philosophical. A Medusoid is a gemma, usually detached as a free zooid, and consisting of a bell-shaped mantle or *umbrella*, fringed with tentacles at its margin, and having a cylindrical process *(manubrium)* depending from the centre of its concavity. The manubrium in its most complete form is itself a polype-like organ, sometimes tentaculated at its free extremity. The umbrella may be regarded as a hemispherical swimming disc developed, round the base of the central structure, after the prevailing type of arrangement in these animals. In its full development it contains canals in its substance radiating from the central or gastric cavity of the manubrium, and opening into a circular canal at the free margin of the bell. The ovaries are usually situated at the points of origin of these canals. The umbrella in some species is converted by the cohesion of its edges into a shut sac, which serves as a marsupium or uterus for the nurture of the ova after impregnation. In the sporosac, which is never detached as a free zooid, the manubrium is generally represented by a central imperforate *columella*, and

* Huxley's Oceanic Hydrozoa (Ray Soc.), p. 137. Allman's terms are *medusæ* and *sporosacs*—the last has an unfortunate similarity to Filippi's term Sporocyst, applied to the gemmiparous tubes of the Trematoda, structures of a totally different import.

the umbrella by an investing sac, in which the canals are imperfect or altogether wanting.

Even in allied species, it is to be noticed, that there is a great diversity in the development of these structures. Thus in *Laomedea geniculata* the reproductive elements are contained in free swimming medusoids with marginal cirrhi, and a central proboscidiform mouth, round the attachment of which the ovarian chambers are situated. In *Laomedea Loveni*,* they assume the form of an inflated bell surrounding the ovary, and fringed at its free edge with minute tentacles ; and they remain during their brief period of life attached to the edge of the horny " ovigerous capsule" characteristic of the group, and there emit the spermatozoa or ova with which they were charged ; after which they wither like blossoms, to be succeeded by a new expansion. In *Laomedea lacerata* the ovarian sac advances to the mouth of the capsule, but instead of a bell-shaped envelope becomes invested merely by a thick gelatinous coat. †
Even in the same species there is sometimes as great a diversity in the opposite sexes ; thus in *L. geniculata* the ova are formed in free swimming medusoids, the spermatozoa in simple cysts, permanently attached.‡

Professor Allman refers to the reproductive sacs of *Tubularia indivisa*, which never become detached, as of

---

* Allman in Annals of Nat. Hist., 3d Ser., IV., p. 140—the same species apparently which Dr. Wright refers to as confounded by Johnston with another having free medusoids, under the name of *L. dichotoma*. (Edin. Philos. Journal, Jany, 1859, p. 110). This form of extracapsular medusoid, Allman terms *meconidium*, from its resemblance to a poppyhead.

† Dr. Wright, Op. Cit.

‡ At least Professor Schultze is quoted by Dr. Wright (Edinburgh Philosophical Journal, July, 1856, p. 85, note) as having discovered sperm capsules in this zoophyte, but there is a degree of confusion among the species which somewhat invalidates the conclusion in the text.

special interest in establishing the relations between medusoids and sporosacs. He describes them as closed capsules, each containing in its interior a second sac, with four canals in its walls, radiating from the base, and communicating with another circular canal round an opening at its apex. An imperforate columella or manubrium projects into the interior from the centre of its concavity, so that we may be said to have all the parts represented which are found in the free medusa, except the marginal tentacles, the eye-spots, and some lesser details.* The genital products are formed in the wall of the manubrium. In *Sertularia* the medusoid conformation is less discernible, and the sacs generally mature and discharge their contents while still within the outer capsule. In *Cordylophora* the only medusoid features presented, either by the spermatic or the ovarian cysts, are the manubrium and some irregular longitudinal canals in the wall of the sac ; there is no aperture (at first) nor any circular canal. In *Hydractinia* we have the columella without the canals ; and the sporosacs of some species of *Plumularia* and *Eudendrium* are if possible of still simpler structure, the latter containing but a single ovum.† The progress of degradation finally reaches its maximum in the common *Hydra*, in which the large medusæ of the "*Hydra Tuba*" are represented only by spermatic and ovarian cysts of the most rudimentary organization, attached to the exterior of the polype.

An equally great variety is presented by the reproductive organs of the allied orders of Calycophorida and Physophorida, ranging from the form of mere sacs to that of free moving bodies (as in *Velella*), precisely resembling medusæ,

---

* Annals of Nat. Hist., 3d Ser., IV., 48.

† In *E. bacciferum*, Allman (Op. Cit.) describes short blind radiating canals. The closely allied species of *Atractylis* detach well formed medusoids. See many interesting details on this head by Dr. T. S. Wright, in Edinb. Philos. Journal, 1856-57-58.

and developing their generative elements subsequent to their liberation. Even in the same species the male and female apparatus are not always alike in structure. Mr. Huxley remarks, that "the female organs in *Athorybia* are sacs containing a single ovum, arrested, as it were, in the first stage towards the medusiform condition, while the male organs become completely medusiform, but are probably not detached. In *Physalia*, on the other hand, the female organs are free swimming medusoids, while the male organs are simply pyriform sacs."*

There is an equally great variety in the mode of origin of the gamomorphic structures—whether medusoids or sporosacs—from the polype stock. In the Hydroid group they spring from papillary outgrowths of the gelatinous basis either of the polypes or of the common axis of the polypidom. When the papilla is prolonged into a fleshy column, it is called by Allman a *blastostyle*, and generally bears a cluster of sporosacs or medusoids. Sometimes the blastostyle and polypes are naked, as in *Clava* and *Tubularia*; but more frequently they are surrounded by a *gonophore*, or involucral capsule, derived from the horny sheath of the polypidom, as in species of *Campanularia*, *Laomedea*, and *Antennularia*. In *Hydractinia* and *Cordylophora* we have gonophores without any blastostyle, the capsule containing a single sessile sporosac. In *Plumularia cristata* the gonophores occur in clusters, in peculiar receptacles formed, as Prof. E. Forbes has shown, by a modification of the ordinary ramuscules of the zoophyte.†

---

* Huxley, Medical Times, XII., 566. English Cyclop. Nat. Hist., I., 38.

† Ann. Nat. Hist., 1st Ser., XIV., 385.

In the homology of these structures, it is the manubrium which appears to be the representative of an ordinary polype, the umbrella being a superadded organ for natatory or other purposes. The simple gonophore may be taken as the homologue of the horny polyp cell, and the blastostyle as a rudimentary polypiferous ramuscule. The compound

The gamomorphic gemmæ, in the case of the hood-eyed Medusæ again, are detached from their polyp stock as a *strobila* or pile of zooids, by a successive transverse fissuring of the body of the polype just below the tentacles. When the segmentation is completed, there is developed round the remaining portion of the body, immediately below the pile of medusoids, a new circle of tentacles, so that when the whole are thrown off the polype is enabled to resume its pristine form and mode of life.*

Such variations as have just been noticed, though perplexing to the systematic zoologist, are especially valuable to the physiologist, as indicating the true relations of the forms which occur in dimorphous species ; and I think they are sufficient fully to bear us out in the conclusion, that both the bare-eyed and the hood-eyed Medusæ are to be considered as gamomorphic zooids, and the polype-stock from which they spring, as the typical form in each case. In the one the orthomorphic form is, as usual, the most conspicuous phase of the species, while in the other it is quite eclipsed by the resulting gamomorphic zooid, which is really a part of itself—a detached and overgrown organ

gonophore is considered by Prof. Allman to represent the concrete cells of the polypes of such a ramuscule. (Edin. Philos. Journal, VII., 294, IX., 111.) Mr. Huxley professes to adopt the term " gonophore" from Dr. Allman, but the two authors do not appear to use the word in the same sense. The latter applies it to the horny vesicle or urn which contains one or more sporosacs or medusæ, and speaks consequently of *medusiferous* gonophores ; the other applies it to the medusoid itself, and speaks of *medusiform* gonophores.

* According to Desor, the strobila or pile of nascent medusoids is formed inside the oral circle of tentacles; at the expense of the proboscidiform mouth, while the figures and description of Dalzell and Reid represent it as external, the original tentacles being elevated on its summit, and a new one formed on the residuary base, when the whole pile has been detached. Does this indicate a specific difference, as suggested by Dr. A. Thomson in the Cyclop. of Anat. and Physiol. (Art. *ovum* p. 22) ?

of its own system. That is to say, as all these structures are admitted to be strictly homologous, the large and elaborately organized Medusæ, instead of ranking as the typical forms of their species, must be reduced to the level of their humbler representatives in the *Hydra fusca*, and regarded as mere spermatic and ovarian appendages, which, like the staminiferous flowers of the *Valisneria*, accomplish their special function after being detached from the parent organism.*

It is in farther support of this view, that in other Lucernariada—including the genus *Lucernaria* itself—no medusoids are thrown off, but the ova and spermatozoa are formed in immediate connection with the polypoid stock, which, as being the only form in these species, must be taken as the characteristic type of the group.

But, besides such analogical arguments, the contrast between the permanence of the polypiform, and the evanescence of the medusiform state, already noticed, as an argument for the typical character of the former, applies with quite as much force to these Lucernarian polypes as to those of the Coryniform and Sertularian groups. The "Hydra Tuba," indeed, is often spoken of as the "larva" of the Medusa, but there is no true analogy between their mutual relations, and those of the vermiform larvæ among the Articulata to the perfect insects. The larva developes but a single insect, and, with the exception of its lifeless integument, is wholly transformed into it, but the "Hydra Tuba," after throwing off a whole pile of Medusæ from its oral extremity, regenerates its original tentaculated mouth, and returns to the common condition of polype life, till circumstances occur to develope anew a similar reproductive process.†

---

* British and Foreign Med. Chir. Review. I., 203-212.
† Steenstrup on Alternation of Generations, p. 6. British and Foreign Medico Chirurgical Review, as above quoted.

§ 4. The conclusion, therefore, to which a fair comparison of the reproductive process in the two cases now cited —the Trematoda and the Polypifera—seems to lead is this, that though in both we certainly have an alternation of forms in the genetic cycle, still we cannot treat them as parallel. It does not seem possible to do so without leaving out of view two points of obvious contrast.

1. That the polype stock has a much greater permanence than the trematode *redia* or precursory zooid, and may throw off many successive broods of medusoids.

2. That the concluding links of the respective series have really nothing in common but the single point of sexual completeness—the medusoids being sometimes of the most rudimentary structure, more comparable to detached organs than to perfect animals. For the parallel indeed, the single character of sexual completeness must be taken to mark the culminating point of the development of the species, however defective the general organization; and this is precisely what some naturalists contend for. Thus Siebold in his concluding note on the Comparative Anatomy of the Acalepha remarks—" However various the developing forms may be, that one must be regarded as the real one, which exists during the development of the testicles and ovaries." So also Dr. A. Thomson—"Some [Hydrozoa] we have been accustomed to see principally in their largest and most developed condition, as Medusæ; others are best known in that polypoid condition, in which they remain for the longest time; but we must regard that condition in which sexual reproduction takes place as the complete one, and this we have seen is in both the Acaleph or Medusa form, while the polype or polypoid state, however permanent it may appear, is to be looked upon as a preparatory stage."*

But the general analogies of living beings surely indicate

---

* Cyclopæd. Anat. and Physiol., Art. Ovum, p. 22.

the completeness of the whole organization, according to the type of the species, as a juster criterion of the culminating point of development than the possession of any one function, however important. It is true that where the individuality is much broken up, there may exist a difficulty in determining what particular zooid does really represent most correctly the typical organization of the species—a difficulty which can, of course, have no place, where, such zooids not being budded off, all the structures characteristic of the species are united as organs of one individual. Yet as it rarely, if ever, happens that alternation prevails so universally in any natural family as wholly to exclude the more common course of reproduction, the cases—whether many or few—in which this function is performed without disturbance of the integrity of the organism, generally furnish us with sufficient data for selecting the truly typical form, and for determining whether it does or does not coincide with that phase of development characterised by the sexual peculiarities.

Now the viviparous Trematoda, in which there is no "alternation"—the ovum being at once developed into the sole form of the species—clearly indicate that in this group the typical form is that represented by the *Distoma*; and this is never dissociated from the sexual phase, though in the great majority of the order the culminating point of development is not attained till after a more or less prolonged course of gemmation in a rudimentary stage.

Among the Polypifera again, we have the case of the common *Hydra* to determine the polype form as the typical, because in that species the only form; and therefore I conceive we must hold it to be the typical form also in the Sertularian and other alternating species, though no longer that of sexual completeness.*

* A much more perplexing fact—if it be really a fact—is the occasional

In the Protomorphic alternation, therefore, of the Trematoda, we speak of the typical organism and its germ-like matrix; in the Gamomorphic alternation of the Polypifera, of the typical organism and its sexual offset. Both the matrix and the offset may assume, indeed, the form of independent beings, but their life is always transitory and provisional, having reference to one common end—the multiplication of the race—though by different means.

development of Medusæ from the egg without passing through the polype stage. It is said to occur in *Pelagia noctiluea* (Krohn), *Œginopsis Mediterranea* (Müller), and *Trachynema ciliatum* (Gegenbaur). Annals of Nat. Hist., 2d Ser., IX., 37, XVII., 285. Huxley in Medical Times, XII., 506.

Professor Huxley, in retaining the term Medusidæ for such cases, remarks—" It is not yet proved that any of them are developed directly from the eggs of similar organisms. I by no means wish to question the great probability of the supposition that those ciliated embryos which were observed by Müller and Gegenbaur to pass directly into the Medusæ, proceeded from the eggs of similar Medusæ. But, I repeat, there is no proof of the fact."—*Oceanic Hydrozoa, p. 21, and note.* I am equally unable to harmonize some of the statements of Mr. Lewis. He cites Professor Kolliker as having " seen the same species of *Campanularia* producing both eggs and Medusæ," and goes on to mention that he himself had " on the *same polypidom* found some of the capsules filled with eggs, and some with Medusæ," and that in different specimens of one species *(Plumularia Myriophyllum)* " dredged at the same time, and from the same place, he found *capsules containing eggs and also Medusæ;* and others—but not on the same polypidom—*containing eggs and polypes, i.e.,* the ciliated gemmules which we know to be the infusorial stage of the polype." These observations—though unfortunately not stated with the fulness and precision necessary on such a point—seem to point either to the conclusion, that the same species may bear sporosacs with true ova, and free medusoids, the latter developed from gemmæ closely resembling ova—or to another result, still more at variance with our received views—viz., that true ova, formed within the capsules (sporosacs ?) of the polype, and actually impregnated, may undergo direct development into Medusæ.—*Sea-side Studies,* p. 296.

Van Beneden, it may be observed, is quoted by Carpenter for the observation that in some Tubularida both attached ovarian cysts and free medusoids are developed, with no difference that can be traced in their respective products. Compar. Physiol., § 539.

The great function of the germinal or protomorphic zooids is the evolution of the embryos of higher forms, of which they serve as budding-stocks ; that of the sexual or gamomorphic zooids is the development of ova and spermatozoa. These ends accomplished, their vitality ceases, while the typical organism, the offspring of the former class, or the parent stock of the latter, as the case may be, has a much more permanent duration, and may go on for a long time in perfect vigour, sending off crop after crop of ova, or of sexual gemmæ, according to its mode of propagation.

§ 5. The distinctness of these varieties of "alternation" is further shown by their occasional co-existence in the same species. Such an association occurs, for instance, in some Cestoid worms. This is apparent from the sketch before given of the development, first of the Cysticercus, and then of the Tænia, from the six-hooked contractile vesicle, discharged from the egg of the latter. The vesicle, as was shown, is first transformed into the " cyst," which represents the Redia or protomorphic zooid of the Trematoda, and then buds off the Tænia-head (the orthomorphic or typical form). From the latter are again derived, by a second gemmation, the cucurbitiform or sexual segments making up the body of the Tapeworm, which represent the gamomorphic zooids of the Polypifera.

This species was selected in illustration, as the best known, and, in some respects, the most characteristic of the order. It certainly exemplifies as well as any the development of the gamomorphic structures, but the gemmation in the precursory or protomorphic stage is a much less conspicuous feature in it than in many other species, from the circumstance that there is ordinarily formed but one gemma, and that this is never detached from the primary cyst, till the latter disappears on the animal entering on its tænioid phase. In the allied species *Cœnurus,* numerous Tænia-heads are budded off from the same cyst, each of which may originate a separate

Tapeworm. In *Echinococcus*, several successive gemmations of simple cysts may occur before Tænia-heads are formed at all; and it is by its cystic growth that the parasite becomes formidable, the resulting Tæniæ being minute three-jointed worms, of no importance on their own account.

Of the two kinds of alternation thus exhibited by the Cestoidea, it is obviously only the protomorphic or cystic that corresponds to the alternation of the Trematoda—a point which seems to have escaped the notice of Professor Van Beneden, as he applies the term *scolex* alike to the protomorphic Redia or Sporocyst of the *Distoma*, and the orthomorphic Tænia-head of the Cestoid worm; and that of *proglottis* to the *Cercaria*, or larval stage of the typical *Distoma*, and to the ovigerous segments budded off in the gamomorphic phase of the Tapeworm. As this application of the nomenclature confounds forms which I cannot but consider of different significance, I have thought it best to avoid using these terms, though fully sensible of the advantage of having short names to express corresponding stages in the cycle of development and reproduction in different species.*

The association of these two kinds of alternation may be traced in a more latent form in the Polyzoa, as will be more fully noticed afterwards. Such co-existence, however, is, on the whole, exceptional, for it would appear that organisms, which are propagated by protomorphic gemmation, do not ordinarily throw off sexual zooids, and that species, in which the latter phenomenon occurs, do

---

* In the use of the term *Strobila* there seems to be an equal ambiguity, as it is sometimes applied to the typical form (as of a polype), from which sexual zooids are eventually to be budded off, sometimes to the aggregate of these zooids, already formed, but still adhering in a chain or pile to the typical stock. Its derivation *(strobilus,* a fir cone) is, of course, suggestive of the latter meaning, and it was to an animal in this transition stage that the name was first applied by Sars. Steenstrup on Alternation, p. 19.

not usually furnish instances of pro-embryonic forms. We may observe indications, indeed, of such a tendency even in the Cestoid order, which has been referred to as furnishing the most satisfactory examples of the association, for we find that when gemmation is very marked in one stage, whether protomorphic, as in *Echinococcus*, or gamomorphic, as in the common Tapeworm, it is in general proportionally in abeyance in the other.

§ 6. There are, however, I conceive, certain cases of alternation, which are not properly referable either to the protomorphic or gamomorphic stages of the genetic cycle, but depend rather on the occurrence of gemmation in the intermediate or orthomorphic stage—that is, during the subsistence of a more fully developed condition of the typical organization, but prior to the maturation of the sexual organs. This I believe to be the case with the alternation which is so striking a feature in the reproduction of the *Aphides*. The remarkable phenomena attending the propagation of these insects is too well known to require any detailed account to be given here, and I shall confine myself therefore to such points as have some bearing on its relation to the other forms of alternation, especially as farther reference will have to be made to the case, in connection with the enquiry into the difference between ova and gemmae.

The eggs of the *Aphis* are laid in autumn, and hatched, as usual, in spring, but the primary larvæ never attain the full completeness of the insect type. They resemble the working bees in neutrality of sex, though certainly not in sterility, as by a process of internal gemmation they produce large broods of young, resembling themselves both in their low grade of development and in their unassisted powers of increase.* But as the larvæ are incapable of

* As this process of multiplication may go on the whole summer for nine or ten generations, an easy calculation will show that many millions of Aphides may spring from a single larva.

withstanding the winter cold, all this increase would not avail to the perpetuation of the race, were it not for the provision that, towards the end of summer, a brood appears capable of complete transformation into perfect insects of both sexes, precisely similar to those of the preceding season, which were the progenitors of the original neuter brood hatched in the spring.

Now, if we compare this case of alternation with the two previously noticed, we cannot but observe that it is more closely allied to that of Trematoda, than to the modification occurring among the Polypifera, for of all the successive forms in the series the one which bears the sexual organs is undoubtedly that which approximates most to the ideal of the insect type.* So much is admitted by the general phraseology applied to the case, all the sterile forms which proceed being, by common consent, termed *larvæ*. But this, at the same time, suggests the suspicion that the gemmation is referable rather to an early stage of the orthomorphic than to the protomorphic or germinal phase of development. For if the zooids are really equivalent to the larvæ of insects generally, their condition corresponds to the embryonic state of other animals—for a larva is simply a naked embryo. Now, the larval or embryonic state decidedly belongs, as a whole, rather to the typical than to the germinal stage of development. For the former must be held to include the whole period, from the first commencement of permanent organization till it attains the standard of the adult. The commencement of the permanent organization is a definite

---

* It seems not to be the case, however, as is sometimes assumed, that the precursory forms are all without wings, and only the terminal males and females furnished with these organs. It is doubtful if the true females are ever winged, but both neuters and males appear to occur in both guises, the determining conditions being yet unknown. It is much the same with the alternating species in the group to which *Chermes* belongs. (Huxley, in Annals of Nat. Hist., 3d Ser., II., 214. Leuckart, in do., IV., 321.)

epoch in the life history; we can readily distinguish a point of time when the cellular germ-mass first presents the incipient traces of that structure which is characteristic of the Vertebrate, Articulate, or other leading type of organization; but we cannot draw such a definite line of demarcation between any two successive periods of the ensuing embryonic development: the subsequent changes are of degree rather than of kind—a gradual, and, on the whole, a continuous unfolding of the perfection of the type from the rudimentary traces which constitute its first commencement. Undoubtedly, we meet occasionally with arrests of development more or less complete and extended —and nowhere more markedly than in the class of insects —but these are all of an irregular or adventitious nature, as is shown by the entire absence, in many cases, of any break in the continuity of the process, and by their variability when they do occur. If, then, we regard embryogeny, or larval development, as constituting the initial state of the typical or orthomorphic stage—but as, on that very account strictly belonging to it—then, we must refer the alternation of the *Aphides* to the commencement of this stage rather than to the protomorphic, because the organization has already acquired that partially advanced development characteristic of the larvæ of other insects, before the process of gemmation comes into play. We cannot say here that the primary product of impregnation buds off a set of embryos of a higher organization; it is rather a larva—that is, a naked embryo—already so far advanced on the insect type, that buds off a series of similar larvæ, the last only of which become perfect insects.*

---

* Dr. Carpenter, while he virtually admits the distinction of the forms of alternation here termed Protomorphic and Gamomorphic, by comparing the zooids of the latter to detached reproductive organs, and by terming the former a process of "larval gemmation," at the same time uses this expression to include the development of the precursory forms both

§ 7. The conclusion, therefore, at which we arrive is this, that each of the three stages into which the life history has been divided (protomorphic, orthomorphic, and gamomorphic), may become the scene of a process of gemmation, attended by an alternation of forms: and, under one or other of the varieties of alternation thus admitted, I am persuaded all the cases known, in both kingdoms of nature, may readily be ranged.

To the protomorphic, I would refer, along with the Trematoda, the Echinodermata among animals, and the Mosses and Hepaticæ among plants; to the orthomorphic, I would refer such Crustacea as *Daphnia* and *Cyclops*, whose reproduction is, to some extent, parallel with that of the *Aphides** ; and to the gamomorphic, the *Salpæ* in the animal, and the Ferns and their allies in the vegetable kingdom.

It is not common, as has been stated, to meet with the concurrence of more than one kind of alternation in the same species. The Cestoid worms, however, have been already mentioned, as furnishing examples of alternation,

of the Trematoda and the *Aphides*. Very probably, indeed, notwithstanding the distinction now indicated, the two modifications may be connected by a series of intermediate gradations. All such precursory forms, being sexless, come under the term "agamozooid,' used by Huxley and Lubbock. See Carpenter's Compar. Physiology, 4th Ed., p. 558, 597, 529.

* The species of the allied family of Coccida, and of the transitional genus *Chermes*, present, according to Leuckart, phenomena somewhat analogous to those occurring in the reproduction of the *Aphides* (Annals of Nat. Hist., 3d Ser., IV., 321). Among the cases of orthomorphic alternation should probably be included also that of the *Pteromalus*, as described by Filippi. Here a caterpillar is developed within a maggot-like larva, which by its growth it reduces to a mere sac, and in due time passes into the pupa state, and comes forth as an insect of the Ichneumon tribe. The case may possibly be one of parasitism, but the uniformity of the occurrence is opposed to this view. See Annals of Nat. Hist., 2d Ser., Vol. IX., 461, and Carpenter's Princip. of Compar. Phys., 4th Ed., 597.

both in the protomorphic and gamomorphic stages ; as also, in a more latent form, the Polyzoa, which will come under consideration again, as illustrating the connection between alternation of generations and certain phenomena occurring in the higher animals.

§ 8. Some Annelida may be said even to present phenomena akin to those of alternation in all the three. The earlier alternations are most readily to be traced in the branchiated species. In these the primary infusorial form in which the young quit the egg may be compared to a protomorphic zooid, such as the "six-hooked embryo" of the Cestoidea. As the latter buds off a tænia-head, so this infusorian gives origin, on one side, to the cephalic, and, on the other, to the caudal and penultimate segments of the Annelidan—conjointly representing the commencement of the typical form—after which the ciliated part gradually wastes away. There now follow a series of successive acts of gemmation, the penultimate segment budding off a new joint on its caudal aspect, which in turn repeats the process, and so on, till there is formed eventually a long train of such segments, each of which (except the tail piece), is derived by gemmation from the one immediately in advance of it. Hence, all the links of this later gemmation are as distinctly referable to the orthomorphic phase as are the successive gemmations of *Aphides*, from which they differ only in continuing in organic union with each other, instead of assuming the position of *free* zooids.

This correspondence is much closer than that which subsists between the Annelidan and the Cestoid worm. Much as these resemble each other in the composite character of the long vermiform body, made up of a series of derivative and semi-independent gemmæ, there are two important points of difference between them.

1. The segments of the Cestoid worm, though of different ages and degrees of maturity, are all originally budded off

from one point—the back part of the Tænia-head—and therefore, except in so far as they have multiplied afterwards by sub-division, may be said to constitute but a single generation, being all the progeny of one parent stock, while the segments of the Annelidan are a succession of generations, each being derived from that immediately in front of it ; so that the hinder ones are the youngest, not, as in the Tænia, those nearest the head.

2. The segments of the Cestoid worm are clearly gamomorphic or sexual gemmæ, for as soon as they show any distinct organization at all, it tends to the development of the reproductive elements. Those again of the Annelidan are, as has been observed, orthomorphic, tending to the completion of the typical body, which attains some considerable elongation before it enters on the sexual stage. As in general it is only the later segments that develope reproductive organs, these indeed may with some reason be considered as constituting a new series of gemmæ, really corresponding to the gamomorphic segments of the Tapeworm. In *Syllis*, *Myrianida*, and some other species, it is obviously so, for the later segments become, as has been already noticed, so many new centres of a budding process, resulting in the formation and detachment of distinct sexual zooids. In this case the primary Annelidan has no sexual organs, but the caudal zooids become gorged with ova, even before actual separation takes place. In the same way other individuals of the species, equally destitute themselves of sexual parts, throw off from their posterior extremity secondary Annelida, with voluminous spermatic organs. When impregnation has been effected, and the ova have attained a certain degree of development, the female zooid becomes completely detached from the parent stock. Its separate existence, however, is of limited duration, for it is ruptured by the growth of the eggs, which are thus dispersed. As M. Quatrefages expresses it, the zooid " is an

animal formed solely to serve as a reproductive machine."*

The zooids, it is to be observed, are not derived from each other, as the rings themselves are, but are the results of a separate budding process in certain of the posterior segments; and the conversion of segments into zooids takes place in the reverse order of their original development—that is, from behind forwards. Though the hindmost joint is the last formed, its derivative Annelidan is the first matured.†

The sequence of alternation here is, therefore, parallel to that occurring in the Polypifera, only there is less difference between the sexual zooids and the parent stock—the annelidan zooids are neither so rudimentary as some of the medusoids, nor do they ever rise so much above the level of the stock as the large hood-eyed Medusæ. Both the parent animal and its offsets have the obvious annelidan structure, yet there are certain minor distinctions between them—such as might even indicate a generic difference, and it has been ascertained that the young produced from the ova resemble, not the immediate parents, but the primary Annelida, in their general conformation, as well as in their want of sex and their tendency to caudal gemmation.‡

Among the Annelida, therefore, we seem warranted in saying that gemmation may occur in all the three stages before specified,§ though it is only as an occasional occur-

* Rambles of a Naturalist, I., 217.
† Hence the facetious remark of Mr. Lewis is hardly in point, that the family inheritance of the parental tail, in passing to the progeny, reverses the law of primogeniture, and like the stock of baby linen, descends always to the youngest. Sea-side Studies, p. 62.
‡ Milne Edwards, Ann. des Sciences Nat., 3d Ser., Zool., tom iii., p. 170.
§ *Terebella*, in which the gemmation of segments from the infusorial germ, and from each other, has been most distinctly observed, presents also, according to Mr. Lewis, the formation and detachment of caudal zooids. Sea-side Studies, 62.

rence and in the gamomorphic stage, that it appears as a case of distinct alternation, because in the two former the gemmæ remain in union with the parent stock, instead of assuming the character of free zooids.

The remarks now made, and those which the course of the argument will yet call for, touch, I think, upon all the cases of alternation in the Animal Kingdom, above referred to, with the exception of the *Salpæ*, to which, therefore, a passing reference may here be made.

§ 9. Notwithstanding that the reproduction of the *Salpæ* was the first case of "alternation" to attract notice, and that it has ever since been regarded as among the most striking instances of it, it is at the same time one concerning the exact relations of which some doubt may be entertained. From the associated forms having, along with perceptible differences in external configuration, such a close similarity in internal structure, and from their locomotive, circulatory, digestive, and nervous systems being so entirely on a par in the scale of organization, it would be hazardous to pronounce one more than the other the typical representative of the species, so that we are in a sort of difficulty to decide whether the alternation is due to the interpolation of a protomorphic zooid—the solitary asexual form which originates directly from the ovum—or of a gamomorphic zooid—the catenated sexual form budded off from the former. In point of organization there is nothing to indicate to us whether the gemmation precedes or follows the orthomorphic stage of development. From all that we can gather from a comparison of the two forms, we might either compare the solitary *Salpa* to the gregariniform phase of Trematode life, and the catenated *Salpæ* derived from it to the brood of *Distomas*—both alike maturing their reproductive structures as simple organs in the continuity of their own bodies—or, on the other hand, we might consider the solitary form as the typical one—elaborated out of the

germ-mass without any breach of structural continuity—
and compare it to the Tænia-head of a Cestoid worm, or the
zoophytic phase of polype-life; in which case the chain of
sexual *Salpæ* would represent simply an aggregation of
generative structures, provided (like the proglottides of the
Tænia or the medusoids of the polype) with accessory
organs of nutrition and locomotion. Either of these com-
parisons perhaps might stand of itself, but the weight of
analogy is much in favour of the latter; the parallel between
the chain of sexual *Salpæ*, the pile of medusoids, the
jointed "body" of the Tapeworm, and the caudal train of
some Annelida—all alike made up of segments furnished
with proper sex-organs, and maturing true ova—is too distinct to be overlooked. In this direction accordingly it is
that we find those naturalists who have made this class
their especial study are the most inclined to look for illustrations. Mr. Huxley, in particular, is very decided in
favour of this view. In his paper in the Philosophical
Transactions on the Anatomy of *Salpa* and *Pyrosoma*, he
characterizes the aggregated zooids as merely highly individualized generative organs,* and in a lecture at the Royal
Institution he ranks among their homologues the medusiform
zooids of *Physalia* and *Velella*, the medusiform organs of
Diphyes and *Tubularia*, and the egg-capsules of *Hydra*.†

In this view, of course, we cannot recognize any example
of protomorphic alternation among the *Salpæ*. It is by no
means clear, however, that all the derivative Tunicata are
on a footing in this respect with the catenated *Salpæ*. According to MM. Löwig and Kölliker, the stellate clusters
of the Botryllida are formed by a sort of fissiparous division of the original germ. In this case their formation
would be parallel to the protomorphic gemmation of the
Trematoda, or as Dr. Carpenter remarks, "to the free gem-

* Phil. Transac. for 1851, p. 541.
† Ann. Nat. Hist., 2d Ser., IX., p. 506.

mation of mosses whilst yet in the confervoid state. The fact, however, is denied by Professor Milne Edwards, who considers that the cluster is formed by subsequent gemmation from the first individual; and the matter remains open for further investigation."*

§ 10. A few remarks may be made, in conclusion, in regard to the cases of "alternation" which occur in the Vegetable Kingdom. Here marked phenomena of this kind are confined to the Cryptogamia. In the lower section of this division—including Algæ, Fungi, and Lichens—instances probably occur of all the forms of alternation above referred to, as will appear from the summary of the reproductive process in these orders, given in Chapter II. But as no one form occupies a preponderating place, and as most of the phenomena of reproduction are still so imperfectly known in these plants, it would be premature to attempt any generalizations as to the character of the process as a whole.

The higher Cryptogamic species fall very naturally into two groups, corresponding to the Muscales and Filicales of Lindley, and distinguished by differences alike in structure, habit, and development. The structure of the reproductive organs themselves (antheridia and archegonia) is very similar in these groups, the diversity, as already noticed, lying in their direct connection in the former with the leafy axis, while in the latter they are contained in detached gemmæ or phytoids. It was remarked, at the same time, that from these relations of the reproductive organs, some botanists would argue a correspondence on the one hand between the structures from which they spring, that is, between the leafy axis of the fern and the stalked capsule of the moss; and on the other, between the bodies into which the spores germinate, *i.e.*, the prothallium and the leafy

---

* Compar. Physiol., 4th Ed., p. 571.

axis of the moss. But a comparison of objects of such *primâ facie* diversity—objects more unlike than even the large *Medusæ*, and the ovarian cysts of the *Hydra*, and what is more, without any intermediate connecting forms— ought not, I conceive, to be adopted, except on the most convincing evidence. But there is no such cogency in the case, if we assume that the interpolation of a derivative form occurs at a later stage of the genetic cycle in ferns than in mosses ; and for this view we have ample warrant, in the analogy of the Animal Kingdom, where we find corresponding differences between the Cestoid and Trematode Entozoa, and between these and the Polypifera ; while among the latter, even nearly allied species differ in this matter of the interpolation of gemmation. Such a view, I submit, is a less tax on our powers of conception than to regard the minute and fugitive capsule of the moss as the equivalent of the perennial and towering stem of the tree fern. And no less real is the contrast between the evanescent prothallium of the fern and the foliaceous axis of the moss, which, humble as is its mode of growth, has yet such permanent vigour of vegetative power, that its pullulations may eventually come to cover a larger space than that overshadowed by many a spreading forest tree ; for the leafy shoots which, year after year, cover the mossy bank with verdure, and send forth each its annual clusters of capsules from the impregnated archegonia, are often all of them the ultimate twigs or branchlets of one original moss-plant, whose primary axis and its immediate ramifications have long since mouldered away, and gone to form the accumulation of soil in which the present shoots vegetate.*

---

* Mr. Jenner, in some respects, supports this view, though his idea of the homologies of these orders differs in some important points from that here adopted. See Edin. Philos. Journal, April, 1856, p. 269; and Annals of Nat. Hist., 2d Ser., xv. 245.

See also the Translation of Radlköfers' Observations on the Function

In mosses, then, as in the Trematoda, the gemmation must be held to occur in a germinal or rudimentary (protomorphic) stage, whereas in ferns, as in the Polypifera, it is delayed till the formation of an axis, in which the general organization reaches its culminating point, though without the development of organs of fructification, which make their appearance only subsequently in the prothallia—derived from it, as the medusoids are from the zoophyte—both being, as it were, supplementary structures, and having for their main end to supply the deficient organs. In the moss the parts of fructification are attached to the axis more persistently, and with less intervention of floral envelopes than in any other plants, while in the fern they are not produced at all, till after the fall of the fern, spore, and the development from it of their matrix by a subsequent process of independent growth. However disguised by this detachment and growth—the prothallium, as the matrix of the spermatic and germinal elements, must remain the real homologue of the floral organ of a phenogamous plant, and the illustration furnished by the floating flower of the *Valisneria* comes in here even more appositely than in the case of the Polypifera.

It will be observed, however, that in thus indicating a relation between ferns and mosses, similar to that subsisting between the Polypifera and Trematoda, arguments of analogy only have been made use of. No reference has been made to transitional forms, such as those which furnish so important a clue in the last mentioned case, for in fact no such forms have yet been recognised among the Cryptogamia, though, as their existence is quite conceivable, their discovery may possibly reward the labours of some future botanist. A case could readily be imagined, for instance,

of Reproduction in the Animal and Vegetable Kingdoms. Annals of Nat. Hist., 2d Ser., xx. pp. 241, 344, 439.

in which the impregnated archegonium of a moss should mature but a single spore, germinating directly into its own protonema and moss, as the egg of the viviparous Trematode is at once developed—without the intervention of any nurse animal—into the likeness of its parent. Or again, we might conceive the "fern-spore," while still attached to the frond, developing a prothallium, which might be compared to the flower of the *Ruscus aculeatus;* such a prothallium would then stand in somewhat the same relation to the original axis, as the spermatic and ovarian sacs do to the body of the common *Hydra*.

Though the discovery of cases like these would no doubt confirm the parallel just indicated, yet, even as the case stands, there seem to be sufficient grounds for assuming the existence among Vegetables, as well as in the Animal Kingdom, of varieties of the so-called alternation of generations, distinguished from each other by the period at which the budding process is interpolated in the cycle of successive forms.

V.

INTERPOLATION OF A CONTINUOUS PULLULATION IN THE GENETIC CYCLE.

§ 1. To complete our survey of the Genetic Cycle in the alternating species, it is necessary to advert more specifically than has yet been done to the fact, that this cycle is at times much extended by a continuous succession of gemmations, all of the same general character. Thus we find that in some of the best marked cases of alternation of generations, the offsets produced in a particular stage of development are not transformed immediately into the next phase in the life-history of the species, but that there occurs a continuous series of zooids [or phytoids], budded off from each other, and all referable to one phase of development, which is succeeded in due time by that next in order, as if there had been but a single intervening link. In these cases, instead of a regular alternation of free gemmæ with true embryos, we find the general propagation of the species effected by gemmæ alone; the sexual zooids, which give origin to the fecundated germs, recurring only at longer or shorter intervals, which are occupied by continuous succession of non-sexual zooids, bearing a general resemblance to each other, though with differences frequently in details.

In certain cases—as among the Trematoda—the links are but few in number, and apparently fixed in each species, and there are generally appreciable differences between the forms which succeed each other in the series; in others we have an indefinite succession of like forms, and the periodic recurrence of the sexual zooids, which terminate the series, seems to depend on circumstances, being a provision to meet contingencies which would otherwise imperil the continuance of the race. Thus in the *Aphides* the male and female insects appear at the close of the season, when it is necessary for the preservation of the species that true ova should be formed, more capable of resisting the winter cold than the rudimentary forms, which are propagated from each other in such quick succession during the summer.

§ 2. Such a process of continuous gemmation may occur, as it would seem, at any period in the life-history of a species. We have an instance of it during the course of germinal development in *Distoma* and *Echinococcus* among animals, and in the successive forms of the capsule and protonema of moss-plants ; and we meet with it also at the other extreme of the cycle, in the medusoid offsets of the Polypifera. Professor E. Forbes makes mention of four modes of gemmation among the Medusæ :—1. From the ovaries, as noticed by Sars in *Thaumantias;* 2. from the peduncular stomach [manubrium] in *Lizzia* [*Cytais*], the gemmæ coming off from each of its four sides in a somewhat symmetrical way ; 3. from the walls of a tubular proboscis into which the manubrium is extended, in a species of *Sarsia*, the gemmæ coming off in an irregular manner from its whole length ; and 4. from the bases or tubercles of the four marginal tentacles, in another species of the same genus.*

* Naked-eyed Medusæ (Ray Soc.), pp. 17, 65, 59, 58. See also English Cyclop. Nat. Hist., I., 25, and Carpenter's Principles of Comp. Physiol., 4th Ed., p. 558.

Dr. T. S. Wright, indeed, regards every medusa as composite, looking on its component members as so many modified polypes. He terms the oral protuberance an alimentary polype, and the contractile marginal filaments and the ovaries, tentacular and sexual polypes, comparing them to the parts fulfilling the same functions in *Hydractinia*—of whose general polype nature there can be no doubt (Ed. N. Phil. Jour., 1857, pp. 315 and seq.) Huxley again regards the whole medusa as a modified polype, and he demonstrates very clearly its agreement in the general plan of structure. The truth seems to be that in proportion as the diffuse

A sort of gamomorphic pullulation may be traced in the very development of the reproductive organs of some medusoids, for while generally in this family the ova and spermatozoa are formed directly in the substance of the walls of the manubrium or central polype, in a few species secondary sacs are budded off as their matrices.* The same idea has suggested itself to Professor Allman, as appears from the following passage in his Notes on the Hydroid Zoophytes :—" In *Laomedea dichotoma, L. geniculata*, &c., the generative elements are never formed in the manubrium of the Medusa bud, but in peculiar bodies situated on the course of the radiating canals. Now these bodies, at least in the Medusa of *L. dichotoma*, which I carefully examined with regard to this point, are constructed precisely on the plan of the sporosacs in *Clava, Hydractinia*, &c. These sporosacs must be viewed as special zooids, representing one term in the 'alternation of generations' of the individual. Just so must the reproductive bodies (sporosacs) which bud from the radiating canals of the Medusa of *Laomedea dichotoma*, be regarded as special zooids, and as representing a term in the life-series of the Zoophyte.

" In *Eudendrium* [*atractylis*] *ramosum*, for example, we

---

vitality of the lower species favours gemmation, in the same degree it tends to confound all the appendages of the body with gemmæ. The detachment of a part may decide its being a gemma (though not absolutely, as we see in the case of the Hectocotylus), but so long as parts continue in adhesion, it must often be impossible to determine between a gemma and a hypertrophied appendage, as all parts tend with development of their organization to take on the typical form of the group, which in the case of the Cœlenterata is that of a bell-shaped polype. See in connection with this the speculations of M'Cosh on the relation between the venation of a leaf—the vegetable unit or phyton—and the ramifications of the corresponding tree. Typical Forms, Bk. II., ch. 2.

* Compare Huxley's account of the development of ova and spermatozoa in the Hydrozoa generally (Oceanic Hydrozoa, pp. 21-22), and Dr. Wright's remarks on the medusoids of certain species of *Laomedea* (Edinburgh Philos. Journal, Jany., 1859, p. 111).

have, therefore, this series represented by two terms—
[ovum] polype, medusa; while in *Laomedea dichotoma* it
is represented by three—[ovum] polype, medusa, sporosac.
In *Eudendrium* the series stops with the production of a
sexual zooid, in the form of a medusa; in *Laomedea* it goes
on through the *non-sexual* medusa-bud, until it finds its termination in the sexual sporosac of the latter."*

But the course of pullulation is most usually interpolated,
after the general typical character of the species has been
first acquired in all respects, save the peculiarities of sex.

§ 3. In fact, the orthomorphic gemmation, already noticed as one form of alternation, almost always runs on into
a continued course of pullulation, the result being either a
swarm of free zooids, as in the case of the *Aphides*, or else
a composite structure, like the polypidom of a zoophyte, or
the leafy stem of a plant. In the last mentioned case, the
seed, derived from the impregnated ovule, emits in germination the primary leaf-shoot of the plant; but, in the majority of instances, before reproductive organs are formed,
this shoot produces leaf-buds, from which other leaf-shoots
are developed, and from these again others originating in
the same way, and so on, till at last that form is acquired
proverbially known as a *vegetation*. As Professor Braun
remarks, " only a small proportion of plants reach the goal of
the metamorphosis (blossom and fruit) in the first generation, the majority attain this term only in the second, third,
fourth, or sometimes not till the fifth generation of sprouts."†

---

* Annals of Nat. Hist., 3d Ser., IV., 368, note.

† Rejuvenescence in Nature, (Henfrey's Translat., Ray Soc.), p. 32.
Braun recognizes three orders of gemmations—Cataphyllary (root and bud-scales, *nieder-blätter*), euphyllary (leaf-shoots, *laub-blätter*), and hypsophyllary (floral shoots, *hoch-blätter*). He holds also that when blossoms
are formed, they occur after a definite number of gemmations, very constant in the same species, and frequently throughout a whole order.

See also Dana in Silliman's American Journal of Science, Nov., 1850,
and the Annals of Nat. Hist., 2d Ser., VII., 318.

At first a general similarity prevails in the successive shoots, amounting in most cases almost to identity—the instances of diversity of leaves on the same tree being quite exceptional—but in by far the majority of plants, the flower-shoots, on which alone reproductive organs are produced, differ widely from the leaf-shoots, though the arguments on which the science of Vegetable Morphology is based prove clearly an essential community of nature between them. In all normal development these floral leaves or *bracts* immediately precede the evolution of the reproductive organs, but the number of successive leaf-buds of the common form seems to be to some degree dependent on external conditions—many plants vegetating vigorously in situations where they never flower.

All that has now been said of such gemmation in plants will apply also to zoophytes among animals, if we merely substitute " polype-bud" for " leaf-bud," and " ovigerous capsule" for " flower-bud."

That the different leaf-shoots in the plant, and the polype shoots in the zoophyte, are so far distinct individuals that they possess a certain independent life of their own, is now so generally admitted, that it is needless to spend time in adducing arguments in its support—the well-known practice of propagating plants by cuttings must occur to every one. The majority of plants, indeed, are equally entitled with zoophytes to be termed compound organisms.*

§ 4. From the gemmæ in such cases so frequently remaining attached to each other and to the parent stock, it is proposed—for want of a better expression—to distinguish

---

* It has been already shown that the compound structure does not necessarily imply the derivation of its component units from each other by gemmation, for it may also depend on their aggregation by a common derivation from the same stock, as in the Catenated *Salpæ*, the pile of Medusæ, derived from a "Hydra Tuba," or the caudal zooids of some Cestoidea. See again in Ch. IX.

the phenomenon by the term *pullulation*, in allusion to the sprouting of leaf-shoots in a tree, which is, in fact, merely a special case of a process of this kind. But the expression is here employed simply to denote a continuous succession of gemmæ in the same phase of development, without restriction to cases in which they remain in this state of adhesion to each other. No such distinction, indeed, could be well carried out ; for gemmæ ordinarily attached sometimes become separate, so as to originate distinct organisms ; and variations in this respect are met with not only in comparing allied species, but even in the same species under altered circumstances.

On the whole, however, such a general rule as this seems to prevail—that in the Vegetable Kingdom, and in the lower divisions of the Animal Kingdom, whose diffuse vitality favours such development, the outgrowths do, with some exceptions, remain in adhesion to form compound organisms, while from the incompatibility of such structures with the more concentrated vitality of the higher animals, such pullulation is either wholly excluded, as in Vertebrata, or, as in Articulata—where it manifests itself exceptionally — the gemmæ are detached as soon as matured, and assume the guise of independent animals. Such an exceptional case is that of the *Aphides* among insects, already referred to. However great the *primâ facie* diversity between the successive swarms of free insects which we here meet with, and the clustered buds of the plant or the zoophyte, the physiological identity of their relations has been well demonstrated by Owen, Carpenter, and others, who clearly show that the only difference lies in a character, the variable and accidental nature of which has just been noticed—viz., the bond of connection between the gemmæ, whose presence in the form of a common axis associates in organic union the successive pullulations of the plant or zoophyte, and whose disappearance in the *Aphides* dissociates the derivative zooids, as the snapping of the thread of a necklace would scatter the beads of which it is made up. "The wingless larval *Aphides* are not very locomotive; they might have been attached to one another by continuity of integument, and each have been fixed to suck the juices of the part of the plant where it was brought forth. The stem of the rose might have been encrusted with a chain of such connected larva, as we see the stem of a *Fucus* encrusted with a chain of connected polypes, and only the last developed winged males and oviparous females might have been set free. The connecting medium might even have permitted a common current of nutriment, contributed to by each individual, to circulate through the whole compound body. But how little of anything essential to the animal would be affected by cutting through this hypothetical connecting and vascular integument, and setting each individual free. If we perform this operation on the compound zoophyte, the detached polype may live and continue its gemmiparous reproduction. This is more certainly and constantly the result in detaching one of the monadiform individuals, which assists in composing the seeming individual whole called " *Volvox globator ;*" and so likewise with the leaf-bud. And this liberation Nature has actually performed for us in the case of the *Aphis*, and she thereby plainly teaches the true value or signification, in morphology, of the connecting links that remain to attach together the different gemmiparous individuals of the *Volvox*, the zoophyte, and the plant."[*]

Though it may be convenient therefore, as proposed by Dr. Lankester,[†] to have appropriate terms to distinguish *attached* gemmæ of all kinds from those *detached* from the parent stock—such as *isozoids* and *isophytoids* for the former, and *allozooids* and *allophytoids* for the latter—we

---

[*] Prof. Owen's Parthenogenesis, p. 60. The correspondence is ingeniously shown by the comparative diagrams in the Frontispiece.

[†] In a communication to the British Association, 1857.

must not allow this terminology to blind us to the accidental nature of the distinction.

§ 5. The repetition of many successive gemmations—or what has been here termed pullation—must be considered a much more variable feature in the reproductive cycle, than the simple alternation of gemmation with sexual generation under any of its modifications; and the extent to which it is carried is still more variable. Thus in the protomorphic alternation of most Cestoid worms, we have but a single primary cyst, which directly originates the Tænia-head; but in the *Echinococcus* there always is one, and frequently are several derivative cysts, interposed before the appearance of the typical form. In the Trematoda again, we have two intermediate forms as a general rule (the infusorial and the gregariniform), but, according to Steenstrup, the latter is occasionally repeated, and that perhaps more than once, before the appearance of the cercariform zooids.* In the

---

* It is this variable multiplication of links which has been the source of ambiguity as to the corresponding forms in the genetic cycle of the Trematoda and Cestoidea respectively. Professor Van Beneden correctly identifies the immediate product of the ovum in each case, under the name of *proscolex* (the "infusorial embryo of the *Distoma*, and the "six-hooked vesicle" of the *Tænia)*, but he is less satisfactory in his identification (under the name of *scolex)* of the forms immediately succeeding—that is, of the Redia of the *Distoma* and the Tænia-head of the *Cysticercus*—because the length of the preliminary pullation differs in these species. If Filippi is correct in stating (Ann. Nat. Hist., 2d Ser., XX., 130) that in some instances the primary infusorian zooid is itself transformed into the redia, rather than that it generates the latter in its interior, in *this* case the comparison would stand, for the infusorian and Redia are as much amalgamated as the "six-hooked vesicle" of the Tænia-egg is with the primary cyst of the *Cysticercus*. But the majority of the species of *Distoma*, which have two links in their preliminary development, should be compared, not with the *Cysticercus*, which has but one —the primary cyst budding off the Tænia-head directly—but with the *Echinococcus scolicipariens* of Küchenmeister, in which the first cyst buds off secondary cysts as the immediate progenitors of the heads. Those *Distomata* which have more than one Redia might in like manner be compared to the *Echinococcus altricipariens* of Küchenmeister, in which

Hepaticæ, among Cryptogamic plants the spores produced by the protomorphic gemma which becomes the capsule, give origin directly to the typical frond, but in mosses we have a second form interposed—viz., the *Protonema*.

In the orthomorphic gemmation the number of links is still more variable, reaching perhaps their maximum among animals in the ten or twelve generations of larval *Aphides*, but in plants running on to an indefinite extent, and probably depending much on external circumstances, while the remarks which have been already made on the proliferous Medusæ show that pullulation, though less common, is not less variable in the gamomorphic stage.

Various other facts might be adduced to show that the processes do not stand at all on the same footing. Thus we find that even in species in which the sexual gemmæ are detached as independent organisms, it is comparatively rare for the non-sexual shoots, which have pullulated from the original typical form, to be detached in the same way.*

We may observe farther, there are numerous cases in

there is a succession of cystic forms before the characteristic heads of the *Echinococcus* are produced. In this view the Scolex of the Cestoid worm —the Tœnia-head—is represented, not by any Trematode Redia, primary or secondary, but by the early condition of the *Distoma* itself, including the cercarian phase through which it passes in the usual course of its development (but not, as it would seem, universally in all the species), and the *proglottides* of the Tapeworm have no farther representations among the Trematoda than what may be furnished by the reproductive organs developed at a later period in the fully formed *Distoma*.

* The exceptions are unimportant, and occur principally in the aberrant cases before noticed among the Articulata, or, again, among phanerogamic plants, as in the deciduous bulbs of *Begonia*, *Allium*, and various Lilies, *Marchantia*, and some other *Hepaticæ*, &c. In a few plants such deciduous buds may occasionally take the place of seeds, especially if the maturation and impregnation of the ovules is prevented by force of circumstances. An inflorescence bearing such bulbs is termed by botanists *viviparous*, though in the received signification of the word it is here misapplied, the only truly viviparous plants being such as the Mangrove, in which the embryo germinates while still attached to the plant.

which the pullulation seems to lie out of the primary cycle of propagation altogether. So it is in plants, with what Braun calls the "inessential sprouts," that is, shoots over and above those which are necessary to the full carrying out of the series of formations up to blossom and fruit. These inessential shoots are mostly formed subsequently to those essential to the cycle, and in some species are very numerous and regular, constituting the great mass of the vegetation, and enabling the plant to rise up in new generations from the same stock, year after year, and thus repeatedly to produce flower and fruit ; or, again, they may become detached, and serve the purpose of dispersing the plant.* We have phenomena of a like kind among animals also, both in the growth of the ramose polypidoms, and in the reversion of the "*Hydra Tuba*" to the gemmation of common polype-buds, after having thrown off swarms of Medusa-buds.†

* Rejuvenescence, p. 36.
† Dalzell's Remarkable Animals of Scotland, Vol. I., ch. 3. Dr. J. Reid's Anat. and Physiol. Researches, pp. 652-656.

VI.

REPRESENTATION OF PROTOMORPHIC ALTERNATION IN THE EMBRYOGENY OF THE HIGHER ANIMALS.

§ 1. In the foregoing remarks traces of reproduction by the co-operation of the sexes have been noticed, in all the leading groups of both kingdoms of nature. This process still continues to exist, side by side with that of gemmation, even in those lower forms of life, in which the latter is much the most conspicuous mode of propagation. Hence the enquiry naturally suggests itself, whether in the higher species, in which sexual generation is the only well-marked agency for the purpose of propagation, any associated phenomena occur which can be considered as corresponding to the gemmation of the inferior classes ; and whether the alternation of the two processes, so striking in some of the lower forms, has any representation in the case of the higher organisms, and of those generally in which naturalists have recorded no change in the forms propagated from each other. In reply, it may be observed that on a careful survey of the phenomena of reproduction in the principal groups of organized beings, indications do present themselves of a general tendency in this direction, as will appear from some illustrations which will now be offered, and which go to show that in the phenomena of embryological development, and of the maturation of the sexual organs, processes occur which have a certain correspondence with those observed in the best marked cases of alternation of generations.

§ 2. We may consider first the phenomena of embryogeny as presenting a certain parallelism with the (protomorphic) gemmation which occurs in some of the lower species (Trematoda, Mosses, &c.) before the period when the typical form is attained.

No point in embryogeny is better established than this, that the first result of impregnation is the formation in the ovum of a cellular mass, from one point of which is subsequently developed a fresh axis of growth, destined for evolution into the organization typical of the species, while the original germ-mass disappears as a distinct structure. The embryo, in short, may be said to be budded off from the primordial germ-mass or "mulberry body," much as the cercariform larva of the *Distoma* is from the gregariniform product of the Trematode ovum.

There are, however, two striking points of diversity which must not be overlooked. In ordinary and normal development the original germ-mass gives rise only to a single embryo, and no separation takes place between them; the later growth appears simply as a more advanced state of the former, which wastes away, *pari passu*, with the growth of the embryo, becoming a mere appendage of the latter, or, it may be, disappearing altogether. In the cases again of alternation of generation, which we have had under review, the immediate product of impregnation gives origin to *numerous* gemmæ, every one of which may acquire the characters of a typical individual of the species. We find also that these gemmæ generally become completely separated from their germ-parent, and assume the form of independent organisms.

§ 3. But although the detachment of the later growth, and its multiplication, give an apparent distinctness to the cases in which they occur, they can scarcely be considered as characters of sufficient constancy to establish, of themselves, an essential diversity of nature between them and the phenomena of ordinary embryogeny; for we shall have to notice instances of alternation in which they no longer present themselves, while they reappear—at least as an occasional abnormality—in the reproduction of the higher species.

As in the former case, we have examples of alternation, where the form immediately developed from the fecundated germ originates but a single gemma, which never becomes detached from its matrix ; so in the latter, a single ovum has been observed to originate two distinct axes of embryonic growth. Cases of a double primitive trace of organization have been met with in the bird's egg, by Dr. Allen Thomson and others, and it is probably in some such way that we may most feasibly account for the origin of what are termed " double monsters."* At all events we have in these, as much as in the best marked cases of alternation of generations, a production of two more or less typical organisms from a single original germ ; for it is now generally agreed that such monstrosities cannot be well explained on any supposition of the fusion of two independent embryos.

This conclusion rests principally on the following considerations :—

1. In all such monsters the duplicated parts are connected together, and derive their vessels from a common trunk ; we never find a face springing out of the chest, legs implanted on the head, or any such mal-position of parts.

2. Double monsters form a continuous series, in which the degrees and modes of deviation from singleness gradually increase, and pass, without any abrupt steps, from the addition of a single ill-developed limb to the nearly complete formation of two perfect beings ; so that no theory can be tenable that will not account for the simpler as well

* Edin. Monthly Journ. Med. Science (1844), IV., pp. 479, 568, 639. See also Vrolik's article on Teratology, in Cyc. Anat. and Phys.

as the more complete instances of duplicity—that cannot explain, for example, the existence of superfluous limbs. As M. Vrolik remarks, "the limbs are mere off-shoots, and are produced at so late a period, that if we could imagine two embryos to come in contact by their shoulders or pelvis, and a fusion of those parts to take place, we should still have to explain how one of them, leaving only an arm or a leg behind him, could for the rest of his substance, head, trunk, and all, wholly disappear."

3. The two monsters are always of the same sex, which we know, from the case of twins, is very far from being a constant rule with associated embryos.

The theory of the furcation of a germ or embryo, originally single, is farther supported by an observation of Valentin's, that an injury inflicted on the caudal extremity of an embryo on the second day was found on the fifth to have produced the rudiments of a double pelvis and four inferior extremities.[*]

Reference may be made also to the observations on the development of the ova of fishes by M. Lereboullet, according to whom, in particular species—as the Pike—the formation of such monstrosities may be determined at pleasure, by placing the eggs in certain conditions unfavourable to development. In this case the blastodermic ridge forms on its surface two tubercles instead of one, and from each of these an embryonic fillet is produced, the farther development of which gives rise to double embryos of various kinds.[†]

The detachment of a portion from the body of the germinal mass has not unfrequently been observed in the embryogeny of the Molluscan Gasteropoda. In this case the isolated segments may become clothed with cilia, and re-

---

[*] Vrolik, Op. Cit.
[†] Annals of Nat. History, 2d Ser., XVI., 19.

tain a proper vitality for some time, but they have never been observed to undergo any farther development. It would appear to be different, however, when the general mass divides by a process of fission into two or more equal segments, as is stated by Agassiz to occur occasionally in the ovum of a species of *Eolis*, for in this case he has observed each segment to develope a distinct embryo of its own."*

§ 4. This multiplication of embryos in the ovum of *Eolis* shows that it is not among monstrosities only that we meet with cases establishing a transition between the phenomena of alternation and those of ordinary embryogeny ; and the evolution of the Polyzoa furnishes us with a farther confirmation of the same, for in some of this group we have as a normal occurrence what may be compared to the formation of a double primitive trace, or a double monster, in the development of the vertebrate ovum. In the Polyzoa, as has been already noticed, the immediate product of the ovum is a ciliated germ-mass, like an infusorial animalcule, from a protrusion of which a pair of polypes are budded off in succession, the phenomena presenting, as Professor Allman observes, " some remarkable analogies, which would seem to bring the whole process of generation and gemmation in these animals within the domain of the so-called Law of Alternation of Generations."† Yet the case is not commonly reckoned one of "alternation," and in fact agrees with it only in the multiplication of the secondary products of development, for in the majority of the group, neither the two first formed gemmæ,

---

* Lectures on Comparative Embryology, quoted by Dr. A. Thomson, in Cyclop. of Anat. Physiol., art. Ovum, p. 24. Such a division of the yolk after segmentation into several (four) globular masses, I have observed myself in the ova of a Nudibranchiate Mollusc, though I have had no opportunity of tracing the development farther.

† Allman's Polyzoa, p. 41 (Ray Soc.)

nor those which afterwards pullulate from them, ever become quite detached; the whole cohering together to form a compound animal—while the original germinal matrix becomes as completely reduced to the condition of a mere appendage of the structures derived from it, as in the case of any vertebrate ovum.

§ 5. For another illustration of a reproductive process, intermediate between the continuous embryogeny, as it may be termed, of the higher species, and "alternation of generations," we may turn to the Cystic Entozoa, which, prior to their most notable change—the transformation into Cestoid worms—present us with a series of successive forms, all referable to the Cystic phase. The typical form —the Tænia-head—is not, as has been shown, produced at once from the egg, but is budded off as an aftergrowth from the cyst which is the primary development. With considerable differences of detail, this general relation is to be traced in all the species. In *Echinococcus* the Tænia-heads are developed in large numbers, by a sort of gemmation, from the lining membrane either of the primary cyst, or of others derived from it by an intervening process of pullulation. In *Cœnurus* also there is a formation of numerous heads from the original cyst; only they are budded off from a special thickening of the membrane, not from the whole interior, and they subsequently become evaginated so as to protrude from its exterior. In both these cases there is an obvious "alternation," attended with a multiplication of the brood, but in *Cysticercus* the germ-mass is resolved into a simple cyst, from which is evolved a single Tænia-head, the whole process having somewhat the character of a continuous development, like the formation of the embryo of a vertebrated animal.* Indeed, not only is the latter formed

---

* A detailed account of the primary development of the vesicle and head of the *Cysticercus* is given by Mr. Rainey in the Philosoph. Transact. for 1857.

much in the same way on the investing or blastodermic layer of the germ-mass, but it gradually becomes invaginated within the amniotic folds of that membrane, somewhat as the head and neck of the *Cysticercus* are within the primary cyst, in the early stage of *its* development. In such cestoid forms as the *Bothriocephalus*, the progressive development is, if possible, of a still more continuous kind, the Tænia-head appearing to be formed from the original contents of the ovum, by as direct a process of organization as is anywhere met with in animal embryogeny, and the act of gemmation not being obviously represented by any one particular event in the course of evolution.

We have, then, in this single order, a series of cases establishing a transition from a well-marked "alternation," attended with multiplication and detachment of gemmæ, to ordinary continuous embryogeny. This argument is of all the greater cogency if *Cœnurus* be, as Siebold contends, a mere variety of *Cysticercus*. Even in admitted forms of *Cysticercus*, however, it is probable such multiple gemmation of Tænia-heads may at times occur, for when the cyst developes itself in the brain (the usual seat of *Cœnurus,)* it occasionally buds off secondary cysts, all of which, it would seem, may bear Tænia-heads.*

§ 6. The Echinodermata afford another interesting illustration of such a transition series. In some species of the class, the primary germ-mass has so much the character of a complete animal as to have been mistaken for one of a totally different family, and named accordingly;† while the new axis of growth, which appears at one point, and increases *pari passu* with the decline of the matrix, has all along such a distinct character, and so much individuality of its own, that nothing but actual observation could im-

---

* Siebold's Memoir is translated in the second volume of the Ray Society's Edition of Küchenmeister's work on Parasites.
† "Pluteus," in *Echinus* and *Ophiura*. "Bipinnaria" in the Starfish.

press the conviction that the one is but a farther advanced stage of the other. Here the majority of naturalists regard the process as a modification of alternation, differing only from that of the Trematoda in the gemmation developing but a single offset from the primary zooid, instead of a numerous brood, each individual of which is in turn equally prolific.*

Yet in this same class there are other species *(Echinaster* and *Holothuria)* in which the organization of the original germ-mass is so little advanced, and that of the typical Echinoderm is so confounded with it, that there is positively less distinction between the two, than exists in the evolution of the higher animals, between the so-called mulberry body, and the proper embryonic organization which subsequently originates from it. Dr. Carpenter, indeed, points out differences in the embryogeny of the Vertebrata, in some degree analogous to those just referred to in the development of the Echinodermata, observing that the umbilical vesicle, which is cast off from the fœtus of the Mammal—like the *Bipinnaria* of the Starfish—is gradually taken—like the *Pluteus* of the *Echinus*—into the body of the Bird; while he finds in the mode of evolution of the Batrachian Reptiles, the whole of whose vitelline mass goes at once to be converted into the embryo, a representation of the type of development occurring in *Echinaster.*†

---

* There may be added one other point of difference—namely, that in all cases, even where there is the greatest extent of change, certain parts of the original structure—particularly the alimentary canal—are incorporated into the new growth.

† Principles of Comparative Physiology, § 555. A practical difficulty which results from this diversity in the development of the Echinodermata, is that it becomes impossible to apply the same terms with perfect accuracy to all the cases, for while there may be little reason to object to *Auricularia* being spoken of as the *larva* of *Holothuria*, we cannot so well apply the same name to the *Bipinnaria* of the Starfish, which, after it has parted from its derivative Echinoderm, continues for some time its

§ 7. Additional illustrations might be cited, but those just given appear sufficient to establish a certain community of nature in all cases of the kind, provided the survey be extended over the whole course of phenomena intervening between two successive returns of the act of digenesis. In all cases it is probable that two successive steps can be distinguished, which may be termed respectively *germinal* and *embryonic*—the later structure, which alone acquires the typical characters of the species, arising from the earlier by a more or less appreciable process of gemmation. In what is called alternation of generations, this gemmation, in the first place, is remarkably definite, the product frequently becoming detached, and acquiring in some sort a distinct individuality; and, in the second place, it is generally multiple, so as apparently to give rise to a new generation of many separate zooids. We find, too, that the gemmation is occasionally repeated many times over, so that a whole race or series of generations seems to originate from a single ovum. But however striking these points of difference may be, the transitional cases just cited indicate that the diversity is less in the essential nature of the function than in its being exaggerated, as it were, in the lower species, so as to diverge from what may be considered, in some sort, a process common to them and the higher—viz., the gemmation, just referred to, of the embryonic structure from the germinal mass. So long as the exaggeration is merely in its distinctness (as in the Echinodermata) the affinity of the process to the normal course of embryogeny is sufficiently apparent. Even when a new element of discrepancy is introduced by multiple gemmation, we can still find a parallel in the embryogeny of the higher species,

own independent life, and may even, for ought that is known to the contrary, be capable of developing another Starfish, by a repetition of the same process of gemmation, for evisceration, which is so readily born by the adult *Holothuria*, can hardly of itself be fatal to the *Bipinnaria*.

though now only as an occasional abnormality. But when the breach is yet farther widened by one or more repetitions of the process of gemmation, we have a result so totally unlike the ordinary course of reproduction in the majority of animals, that it is with some difficulty we can realize the existence of any community between them. The case seems to stand thus: Gemmation of the embryonic structure may be said to occur in all cases, though in the higher animals only in a latent form, and without trenching on the unity of the organism; while multiple gemmation exists in the higher only as a monstrosity, and a repetition of it is wholly unknown. Still as the difference is due to a progressive exaggeration of a feature common to all, we must regard it as one rather of degree than of kind, and therefore not such as to interfere with a certain essential though hidden community of nature—and this even when we come to apply it to such cases as the series of precursory zooids among the Trematode Entozoa.

This way of viewing the case has a bearing on the question of "zoological individuality," before referred to, for as we consider the germinal mass and the embryo which succeeds it in the light not of two animals, but of mere successive stages in the existence of one and the same animal, so the whole succession of rudimentary Trematoda or *Echinococci* may also, in a certain [embryological] sense, be considered as making up but one animal with the typical *Distoma* or *Tænia* in which the series terminates. The separate links of the chain of succession, at all events, have not the same distinctness of organic individuality in cases of alternation, as belongs to the consecutive generations of animals high in the scale of organization, in whose development the embryonic stage—when the typical conformation begins to take shape—is never broken off from the germinal in any similar way.

VII.

REPRESENTATION OF THE OTHER FORMS OF ALTERNATION IN THE HIGHER ANIMALS, ESPECIALLY IN CONNECTION WITH THE MATURATION OF SEX.

§ 1. REASONS have been given in the last Chapter, for drawing a parallel between protomorphic alternation in the lower animals, and a particular feature in the embryogeny of those higher in the scale—viz., the implantation of the rudiments of the typical organization on the blastodermic membrane, which after cleavage is formed on the surface of the vitelline mass. On similar grounds of analogy we might look for some representation of the other forms in the period of typical development, and in connection with the evolution of the sexual organs.

The same analogies, however, would lead us to expect that the traces of such phenomena would be far from evident in the former. Even in the lower animals, as we have seen, gemmation in the orthomorphic stage rarely puts on the form of alternation, because the gemmæ seldom separate as free zooids, generally remaining in adhesion to the common stock, and going along with it and with each other to make up a more or less composite structure. We find further, that, as we ascend to the higher forms, the individuality of the associated parts becomes always less marked, the zooids assuming more and more the position of mere organs, and the budding process gradually merging into one of continuous growth. In the higher animals, indeed, any indications of gemmation, in the orthomorphic stage as distinct from growth, are very equivocal; possibly, however, we may trace them in such phenomena as the renewal of the hair, teeth, antlers, and other cutaneous appendages, or of the whole integument in certain species—the limited reproduction of lost parts—the abnormal sprouting of supernumerary limbs—and the still rarer monstrosity termed *fœtus in fœtu*. Perhaps, too, some of the normal structures of fœtal life may be considered as early orthomorphic or larval gemmæ; *e.g.*, the placenta of Mammalia, the allantois, the branchial tufts of Batrachia, and the ciliated lobes of Gasteropoda.

§ 2. We come next to enquire into the phenomena which, in the higher animals, may be regarded as standing in the place of the variety of alternation depending on gemmation in the gamomorphic stage. To this a certain analogy has just been indicated in the processes connected with the development of the reproductive elements, which may be held to represent this modification of alternation, somewhat in the same way that the succession of forms in embryogeny shadow out the budding off of zooids in the rudimentary or protomorphic stage.

Some of the points which suggest such a correspondence will now be briefly noticed.

§ 3. To begin with one of minor importance, it may be observed that the periodicity which so commonly characterizes the development of the sexual organs seems to assert for them, even in the higher species, a relation to the body at large somewhat different from that held by the other members. Much in the same way, for instance, that the "Hydra Tuba," after a certain lapse of polype life, matures a succession of reproductive structures, in the guise of medusa-buds, and on their detachment relapses again into the ordinary polype life, till other periods of reproductive activity occur, marked by similar phenomena; so we find that among animals generally the parts ministering to the generative function have regularly recurring periods of increased action, dependent on the development or gemmation, as it were, in their substance, of swarms of cells, bearing the reproductive corpuscles, and comparable in a degree to the piles of medusoids budded off from the polype stock. During the intervals the function is commonly entirely suppressed, and the parts, so largely developed during the period of action, are in some cases not to be distinguished at all. Even in species characterized apparently by continuous fertility, the phenomena of ovulation in the female, which are always more or less periodic, are generally admitted to indicate corresponding differences in the proclivity for conception.

Another argument in the same direction, and one probably of greater force, may be drawn from the lateness of the development of the reproductive organs. In some species they would seem actually not to be formed at all till the breeding period arrives. This has been already noticed in the case of the *Distoma*, in which the appearance of these organs constitutes almost as marked an epoch, as the previous budding off of the *Cercariæ* from the interior of their gregariniform matrix. In some instances it is equally dependent on the transference of the parasite to an animal of different species from its former victim. The *Cercaria* which has quitted the snail, on its discharge from its parent sac, for the body of an aquatic insect, may in this position complete its larval development, and yet may not mature its sexual organs, till, by its new host falling a prey to the voracity of some bird, it is transferred to the intestine of a warm-blooded animal.

Another peculiarity in the development of the sexual organs may be noticed in this connection—viz., that among insects which live in communities—such as Bees, Ants, and Termites—these organs are never perfectly developed at all, except in a few individuals, the majority remaining neuter in function, if not in all cases absolutely devoid of the structural characters of sex.

Even, however, when such organs are normally matured in all the individuals of the species, and when, as is generally the case, traces of them do make their first appearance along with the neighbouring viscera,* it is still a constant rule that they lie in a latent condition long after the full development of other parts, and are the last to take on a state of functional activity. This only occurs on the formation in their substance, at the proper period, of numerous cells containing the peculiar reproductive corpuscules (spermatozoa and ova); and it is not till the development of these, that the bodies in question, which up to this time have been mere masses of parenchyma, can rightly be called organs of reproduction. Hence probably the difference between the common case, and that of the species in which these organs appear to be formed *pro re nata*, may not amount to more than this, that in the former the gemmation extends only to the cells producing reproductive corpuscules, while in the latter it embraces also the general envelope or viscus in which they are contained. Or perhaps in propriety we ought to make such a distinction between these formative cells and the stroma in which they lie, as to regard the latter as merely a proliferous tract of the parent's body—like the base of the Tænia-head, or the capsule of the Sertularian polypidom, or like the " funiculus" of the Polyzoon presently to be noticed—and to consider

---

* This, however, is probably conceding too much, for it is now well ascertained that the generative glands are not part of the organization as originally laid down, at least in Mammalia and Birds, but that the large masses occupying each side of the abdomen of the embryo at an early stage of development, which are known as the Wolffian bodies, and have themselves but a provisional part to fulfil, form the primordial matrices, upon which both the urinary and genital organs are developed. See a farther reference to this point in the concluding Chapter.

the former, that is, the sperm-sacs and ovisacs, which are in due time developed in this matrix, as the true organs of reproduction. In this view the medusoids of the Polypifera, and other such sexual zooids, would differ in themselves from the organs of reproduction of the higher species only by the greater complexity of their internal structure, and the possession of certain appendages, required for their maintaining an independent existence.

§ 4. In support of such a relationship reference may be made to the reproductive organs of the Polyzoa, as occupying a position intermediate in some respects between that of derivative gemmæ, and that of mere organs of the particular polypes in connection with which they are developed. Such at least is the view of Professor Allman. " If the formation of the ovary," he remarks, " be attended to, it will be seen that this body is developed at a later period from the walls of the original sac-like embryo, which have undergone slight changes and have become the endocyst [inner coat] of the more mature Polyzoon, and it will be at once perceived that this development of the ovary takes place in a way which may obviously be compared with the formation of a bud, that—at least in *Alcyonella*—it occupies exactly the position in certain cells that the buds destined to become polypides [polypes] do in others, and that at an early stage of polypide and ovary it is scarcely possible to distinguish one from the other ; so that the idea is immediately suggested that the body here called ovary is itself a distinct zooid, in which the whole organization becomes so completely subordinate to the reproductive function as to be entirely masked, and apparently replaced by the generative organs. This would then constitute a third zooid, which would therefore be a sexual zooid ; it is, however, unisexual (female). Again, we find that upon the *funiculus* [or suspensory ligament of the stomach] there is developed the mass described as testis. Now, if we view this mass as a mere organ of the polypide, we must regard the latter as the second or male zooid; but the testis may perhaps be more correctly considered, like the ovary, as a distinct sexual bud, having the generative system so enormously predominant as to overrule and replace all the rest of the organization ; this bud, like the ovary-bud, being also unisexual, but with a male function."* In support of this view he points out that certain undoubted buds, known by the name of *statoblasts*, are actually at times produced in the same species by this very part—the funiculus—while in another species the testis is developed more in the situation of the ovary, *i.e.*, from the endocyst. As instances of such a dominant development of the generative system as to supplant that of the other organs, he quotes the males of some Rotifera *(Asplanchna)*; and probably still more striking illustrations might be found among the Cirrhipedes. Dr. Allman observes in conclusion, that in this view " the complete comprehension of a Polyzoon will involve the conception of a ciliated sac-like embryo as a starting point, and a series of buds, of which the last term will consist of a pair of sexual buds, the others being non-sexual ; from the sexual buds a new embryo, like the first, is again produced, which affords a point of departure for another similar cycle . . . analogies, which would seem to bring the whole series of generation and gemmation within the domain of the so-called 'Law of Alternation of Generations.' "† According to this view a twofold interpolation of gemmation must be recognized in the Polyzoa, first in the protomorphic, and again in the gamomorphic stage—and this over and above the continued pullulation of the polypes, throughout the duration of the typical or orthomorphic phase, by which the ramified structure is formed, known as the polypary or polypidom. It is to this sequence Allman

* Allman, Op. Cit., p. 41.   † Op. Cit., pp. 41-12.

seems to refer above, when he calls one of the generative organs " a third zooid."

§ 5. Another illustration of the semi-independent position of the sexual organs is offered by some aberrant forms of the Crustacean type, especially by species of Cirrhipedes and Lerneans, which undergo what is termed a retrograde metamorphosis. These species commence life as active Entomostraca, but afterwards lose the power of locomotion, attaching themselves to the integuments of marine animals or to other bodies, while at the same time they become variously transformed—or deformed, as we should say, in the majority of cases—by the obliteration of certain parts of the typical organization, and the disproportionate development of the reproductive organs. Here the nature of the change is such as to suggest doubts of the correctness of the common view of the Entomostraceous stage being a larva state, and the subsequent transformation, though of a retrograde kind, one analogous to the metamorphosis of an insect. Another view at least *may* be taken—namely, that the transformation is a sort of gamomorphic alternation, the reproductive structures in many respects standing more in the position of superadded sexual zooids than of constituent organs. Prof. Owen ranks the case as one decidedly of the nature of metagenesis,* and there is at all events a much more complete change in the relations of the parts than in any case of undoubted metamorphosis.†
The greater tendency of the males to retain their original

---

* Parthenogenesis, p. 25.

† The only cases in Insects which are at all similar are those before referred to in the gregarious species, where the procreation of ova for the whole community devolves on a single fertile female or queen. The ovaries may then attain an enormous size, as happens most remarkably in the queen of the Termites, whose abdomen at the time she commences laying is from 1500 to 2000 times larger than the rest of the body, the eggs being subsequently deposited at the rate of 80,000 a day for several successive weeks.

form is nowise opposed to this view, for we have seen a corresponding diversity in some polypes which form their ova in free well-developed medusoids, but the spermatozoa in simple attached cysts.

§ 6. From such cases we pass by an easy transition to the Cestoid Entozoa with voluminous caudal segments, of later growth, containing the reproductive organs. In the foregoing remarks these have been unhesitatingly considered as gamomorphic zooids—a view adopted by Van Beneden and others of the first authority in helminthology, and one, as it would seem, fully borne out by their disproportionate development, and their ultimate detachment while still retaining their vitality.* At the same time it must be admitted that they may also be regarded as being, for a time at least, integral members of the vermiform body, from their connection with each other and with the "head," by two sets of so-called vascular canals and by continuity of tissue generally.

§ 7. It is obvious that if this gradation through the Polyzoa, Lerneada, and Cestoidea, be admitted as showing a community of nature between the ordinary development of the reproductive organs in the majority of animals, and well-marked cases of alternation of generations, we must include in our generalization all the gamomorphic zooids falling under the latter head ; so that even the large hood-eyed Medusæ must be regarded as representing the reproductive organs of the small Lucernarian Polypes from which they have been budded off—not being after all much more disproportionate in size than the voluminous caudal segments of a large Tapeworm, in comparison with its minute denticulated head.

The same conclusion, indeed, as has been already shown,

---

* Küchenmeister describes the joints as depositing their ova as they wriggle along with a sort of spontaneous movement.

may be arrived at by a comparison of the different modifications of reproductive apparatus occurring among the Polypifera themselves, as in this single group we find a transition, and that even among closely allied species, from cases in which the structures ministering to reproduction are integral parts of the system, and organs of very simple form, to others in which they appear as detached zooids, having the characters of distinct and well organized animals. A gradation is here presented to us of the most continuous kind, from the spermatic and ovarian follicles of the *Hydra fusca* to the Medusæ just mentioned, whose isolation from the parent stock, free powers of locomotion, colossal size, and development of organization, appear almost to raise them to a higher type of animal life. As Professor Huxley remarks, " in passing from *Hydra* to *Rhizostoma*, we thus see the reproductive organs acquiring a greater and greater relative mass, when compared with the organism from which they spring, and, as it were, grouping round themselves and subordinating to their own perfection a greater and greater number of morphological elements. First, they are parts of the body wall, indistinguishable in form from the rest ; then they are distinct sacs ; then they are sacs with a *gonocalyx* [bell-shaped envelope] ; then that gonocalyx becomes a well developed contractile organ; next the reproductive apparatus is detached, and swims about independently by means of its gonocalyx or umbrella ; and, finally, it acquires total independence, feeding and nourishing itself, and attaining the most complex organization exhibited by the class to which its originator belongs."*

§ 8. Granting, however, on the grounds now alleged, that a general community of nature may be admitted between the reproductive organs of the higher animals, and the gamomorphic zooids of the lower species in which alter-

---

* Oceanic Hydrozoa, p. 18.

nation prevails, still the great amount of apparent diversity seems to call for some enquiry into the causes to which this may be due. These appear to be mainly the two following:—

1. The *detachment* of the parts of reproduction from the organism producing them, much in the same way that the typical form is separated as a free zooid from the original germ-mass, in the protomorphic form of alternation.

2. The development, in connection with these parts of reproduction, of accessory organs of alimentation, locomotion, &c.

From the influence of these processes on the general organization, the phenomena of reproduction are, in many cases, made to assume a character as unlike that of the higher animals as can well be conceived; but the instances of transition forms just noticed are sufficient to show that differences in these points cannot be allowed any weight in determining for or against their essential community of nature.

In regard particularly to what may be termed the adventitious organization of the reproductive zooids, as compared with mere organs fulfilling the same function, this conclusion is strengthened by the contrast of phenomena of an opposite kind—the degradation of individuals in certain species to the position of mere sexual mechanisms. Wonderful, as it undoubtedly is, to find the homologues of the reproductive organs at times so elaborately organized as to present the characters of distinct animals, and even to simulate species higher in the scale than those of which they are really but detached members, it is not perhaps more wonderful than to meet with animals of undoubted zoological distinctness—as the males of some Rotifera and Cirripedia—whose organization seems almost limited to the parts required for the performance of the sexual function.*

---

* From the observations of Mr. Gosse in the Phil. Trans. for 1857, it

In some of the latter the animal certainly has more the appearance of a mere detached organ, being, by defect of the apparatus of alimentation and locomotion, reduced to the condition of a minute spermatic cyst. Yet its status as a distinct animal is established by its having the same origin from the impregnated ovum, and passing through the same intermediate larva form, as its more highly organized mate; as well as by the gradation through different species from this extreme of degradation up to the entire fulness of the type of the order.*

In the structures now contrasted—the so-called parasitic males, on the one hand, which have just been noticed, and the large Medusæ on the other—we have examples of the two extremes of organization; in the one case we have a *member*, organized above par, so as to simulate a complete

would seem that an absence of alimentary organs is a general feature in the males of the Rotifera—he has found it in all he has ascertained, or has reason to believe, to stand in this relation. The first observations were those made on the genus *Asplanchna*. See Brightwell, in Ann. Nat. Hist., N.S., II., 154; Dalrymple Phil. Trans., 1849, p. 340; and Gosse An. Nat. Hist., N.S., 2d Ser., 18.

* In *Ibla* there is a certain amount of embryonic or rudimentary organization; some species of *Scalpellum* have the ordinary structure of the class, but others are extremely rudimentary, and when mature "may be said to be essentially mere bags of spermatozoa," having no mouth, or other viscera, but those of reproduction. The males of *Cryptophialus* and *Alcippe* are, if possible, still more rudimentary, for they are reduced to an outer envelope, a single eye, a spermatic gland and vesicle, and an intromittent organ eight or nine times their own length, coiled up like a great worm; there is neither mouth, stomach, throat, abdomen, nor cirri. All these rudimentary males are microscopic, sometimes not exceeding the size of one of the ova which they have to impregnate. They are in some cases attached in numbers at the same time—as many as 14 in *Alcippe*—and their existence being transitory, fresh relays attach themselves as the ova become ready successively for impregnation. See Darwin's Cirripedia (Ray Soc.), I., 291, II., 26, 561, 586; Carpenter's Comparative Physiology, 4th Ed., p. 606. Some prior observations of Mr. Goodsir of a similar import will be found in the Edinb. Philos. Journal for 1843, p. 83, XXXV.

animal ; in the other we have a true *animal*, so far below par in its structural development as to resemble a mere organ. The contrast shows in a striking way that the suppression of normal parts in an animal, or the development of adventitious structures in connection with any particular organ, are not of essential importance in determining what has been termed by some authors " zoological individuality."

The other character—the detachment of the sexual structure—is not, as it would seem, of more importance than its amount of organization, allied species both of Polypifera and of Cestoid worms differing among themselves in this respect. It is even said that in some Tubularian polypes there occur in the same species both sporosacs and medusoids.*

It may be mentioned, too, that even so high in the scale of organization as among the Decapod Crustaceans, we meet with something analogous to the elaboration of the reproductive corpuscules after their envelope has become detached, for in the case of certain crabs it would seem that the spermatozoa are not fully matured till after the vesicles in which they are formed have been introduced into the spermatheca of the female.†

In fact the whole question of detachment hinges on the proportionate development of the *somatic* life, *i.e.*, the life of the body as one whole, and the more or less independent life of its several organs, or what we may term the *topical* or regional life. In the higher animals the special actions of the several organs are as completely subordinated to that of the body as a whole, as are the powers of local corporations to the central government in any well-ordered state, yet there still remains sufficient evidence of the real exist-

---

* Carpenter's Comparat. Physiol., 4th Ed., p. 552.
† Goodsir's Anatomical and Pathological Observations, p. 39.

ence of a *distinct* topical life. The hairs and teeth of animals generally, and the antlers of the deer, have already been cited as furnishing illustrations of it. The first set of teeth, for instance, are formed each in its own capsule by a process of local growth, quite independent of that of the neighbouring tissues, nay, in so far opposed to it, that at a certain stage of development the integuments of the gum are partially disintegrated to allow of their eruption. A tooth, thus generated by independent growth, some time after attaining maturity, undergoes a process of decay, ending in its ultimate removal, when a new tooth of the second dentition takes its place by a similar process of local growth. In its turn this tooth also is shed, and though in most species it has no successor, yet in a few there is a constant succession during the whole lifetime of the animal; and this is the general rule in the case of the hair.* Hence in such local formations as teeth, hair, &c., we have, in the way they are marked off from the neighbouring parts, and in this succession of growth, maturation, and decay—repeated again and again, and epitomizing, as it were, the life of the animal on which they grow—evidence of a vitality, quite as defined perhaps in itself as that presented by the free zooids of the lower species, though their functional dependence on the common circulation, and the mechanical bond of a common integument, prevent their exhibiting the more obvious phenomena of a separate life. But as we descend in the scale of organization we come to species, where, from the absence of centralizing influences, the several organs—which are possessed of a vitality, less energetic perhaps, but more enduring than in the higher—become emancipated, as it were, from the control of the general system, and appear as zooids, that is, in the guise

---

* Paget's Lectures on Surgical Pathology. Kirkes' Handbook of Physiology, Ch. X.

of independent beings, rather than as integral parts of the same animal—suggesting a comparison to a loose confederation of Indian tribes, or to the feudal system of the middle ages, rather than to a well-ordered polity of our own day.*

And though the proper organs of reproduction, from their partial independence even in the higher animals, seem, as we might expect, to manifest most clearly this emancipation from the controlling influence of somatic life, yet it is seen very distinctly in others also, as, for instance, in the peculiarly modified tentacle of the *Argonauta*, which, when filled with spermatic fluid, is detached from the body, and finds its way spontaneously to the female for the purpose of impregnation.† The organs of alimentation in the Polypifera may also be considered as an illustration in point. It is quite a tenable view of the compound polypes to regard the whole polypidom as one body, of which the several polypes are organs, combining the characters of mouth and stomach; yet such is their independence of each other, and of the whole group of associated organs, that the play of their functions is not arrested by their separation, it being indeed in some cases the normal course for a polype to detach itself from the rest and become the nucleus of a new community. This relation is suggested especially by such cases as that of *Hydractinia*, in which the various appen-

---

* Some appropriate remarks on the general question here referred to will be found in Professor Laycock's recent work on "Mind and Brain," Vol. II., Ch. VIII.

† The worm-like appearance led at first to its being described as a parasite of this organ under the term of *Hectocotylus* ; and even after its sexual relations were determined by Kölliker, it was still considered as an integral, though rudimentary animal, and in this point of view was employed by Darwin (in the first vol. of his monograph of the Cirrhipedes) in illustration of the nature and relations of the minute parasitic males occurring in certain genera of that group. The discovery of its true nature as a mere tentacle of a Cuttlefish is due to Verany and H. Müller.

dages—which we may term either organs or zooids—are adapted to fulfil different functions for the benefit of the whole—the outer polypes being metamorphosed into tentacular organs of offence and defence, the inner into stomachs, and certain others into spermatic or ovarian cysts.* The same remark will apply to many of the compound Physophorida and Calycophorida.

With such cases before us, it seems very clear that the *detachment* of reproductive zooids, as in one form of alternation of generations, is a point even of less importance than their elaborate organization, in marking them out as of a different nature from the reproductive organs of the higher animals; for when we come to species in which the organs of alimentation, prehension, &c., acquire a sort of independent vitality, it is only to be expected that those of reproduction should also appear as free zooids.†

§ 9. If from the Animal we turn to the Vegetable Kingdom, we shall find that there likewise a correspondence may be traced between certain processes in the fructification of the Phanerogamia and the alternation of the ferns and allied Cryptogamia.

---

* Dr. T. S. Wright, Ed. N. Phil. Jour., April, 1857, p. 305. See also a former note in Chapter V., 2.

† The view here contended for—viz., that the gamomorphic zooids are merely reproductive organs, isolated and hyper-organized into the similitude of distinct animals, may appear to be opposed by certain facts in the alternation of *Syllis* and allied Annelida. Dr. A. Thomson quotes observations of Leuckart and Schultze to the effect that the parent annelidan arrives itself eventually at sexual perfection, after having given off a number of sexual zooids by the caudal gemmation. (Note to Article on the "Ovum" in the Cyclopædia of Anatomy and Physiology, p. 33). As the animal, therefore, comes ultimately to possess reproductive organs within ts own body, we have not, it may be said, to seek for them in the free zooids. Not knowing the exact details, I can only suggest that if the formation of these organs is preceded by a process of gemmation from a local centre, as in the case of the *detached zooids*, then the *region* of the body in which they occur may fairly rank as one of the series, which remains *attached* simply because it is the last formed.

The development of the pollen in the anther, and the embryosac in the ovule of the former division, may be shown to present a distinct analogy in essential points with the antherozoids and archegonial corpuscules in the prothallium of the latter, that is, in the cellular mass which is first formed in the germination of the spore. The resemblance, it must be admitted, does not lie on the surface; the short summary given in the second Chapter shows but little *primâ facie* similarity between the cases, but it exists not the less in those points which are of most essential importance in the reproductive process, and a closer comparison will indicate many points of analogy.

Thus, starting with the assumption that the spore-bearing capsule of the fern corresponds to the ovuliferous carpel of a flowering plant, it is interesting to observe that they are both modifications of leaves; though this is a subject on which it is unnecessary to enlarge here, as full reference has already been made to the circinate vernation and other points of relation between the spore-cases and the ordinary leaves of ferns, in comparing these plants with mosses. The connection of the capsules with the fronds has a parallel in the case of the *Ruscus aculeatus*. But the principal analogies of the reproductive process lie in the changes occurring in the germination of the spore itself, and in the transformation of the ovule into the seed. The main feature of this development is that in the cellular substance of the ovule, or the cellular outgrowth of the spore (as the case may be) minute capsules are generated,—known respectively as embryosac and archegonium,—within which peculiar corpuscules are formed, fitted to become the subjects of impregnation. These corpuscules, which, from the testimony of some able observers, appear to have at first no distinct boundary walls, are converted by fertilization into true cells, and afterwards by a process of endogenous growth into clusters of cells. The part of the cell-mass first formed has but a temporary existence, and is known as the suspensor; that subsequently generated is the true embryo, which in the course of development protrudes an extension downwards (the radicle), and one directed upwards (the plumule).

The particulars now referred to appear to be the essential points in the process, and there is undoubtedly a close correspondence in them, between the two kinds of plants. The differences, which at first appear so great, seem to be reducible to the following:—

1. The shedding of the fern-spores before any of the cellullar growths are formed—and, of course, before impregnation—while the phanerogamic ovule continues *in situ* till the embryo is matured.

2. The development from the spore, in its primary germination, of an external cellular growth or prothallium, which afterwards simulates a cotyledon—whereas generally in the ovule no cellular body is formed distinct from the common tissue of the nucleus.

3. The formation of numerous archegonia in the substance of the prothallium, each with a germinal corpuscule in its interior—whereas in the ovule there is normally but one such capsule—viz., the embryo-sac.

4. The frequent co-existence of spermatic capsules and of archegonia in the Cryptogamia within the same prothallium; whereas in phanerogamic plants, the organs containing the corresponding elements—*i.e.*, the anthers and germens—are from the very first quite distinct structures.

To estimate the importance of these diversities they will require to be considered *seriatim*.

That mentioned first—the premature dispersion of the seed-like bodies—is certainly a very constant point, but it cannot be held to be of a kind to destroy the general correspondence of the course of phenomena in the two cases, for it is evidently a phenomenon of the same nature as the detachment of gemmæ, and like it connected with the more diffused vitality of the lower organisms, owing to which the spore has sufficient intrinsic plastic energy to develope its derivative cellular mass, even when severed from that continuity with the parent stock, which seems absolutely essential to the maturation of the ovule.

The second differential character, on the other hand—that of the great development of the cellular mass during the germination of the spore, so as to form an external prothallium—is not a constant one; for, though it is well marked in the ferns and Equisetaceæ, it is absent in the allied cryptogamic orders of Lycopodiaceæ and Rhizocarpeæ. In these orders the general course of phenomena is very similar, only the cellular mass is never so much developed as to rupture the spore-coat and appear externally, but is limited to a stratum lying immediately beneath that portion of the investing membrane of the spore which is perforated by a sort of micropyle. Hence the term *endothalloid* has been applied to these spores, and that of *exothalloid* to those of the true ferns.* Neither again is the absence of a prothallial structure universal among the Phanerogamia, for in the ovule of the Coniferæ a peculiar mass of cellular tissue of a very analogous kind—the *albuminous body*—is formed within the nucleus, but, like the prothallium of the Rhizocarp, it never protrudes externally. The ovule of the Coniferæ being naked, like the spore of the Cryptogamia at the time of impregnation, the additional protection of a prothallial structure may possibly be required as a substitute for the germen, which encloses the ovules in the higher Phanerogamia till the embryo is fully formed.

A similar inconstancy attaches to the third point of difference—the multiplication of archegonia in the prothallium

* Jenner in Ed. New Phil. Jour., III., 279 (April, 1856).

—as is illustrated in the same natural order, for within the prothallial formation, or "albuminous body," above noticed, there occur in the ovules of the Coniferæ three or four capsular cavities—the *corpuscula*—intermediate in their characters between the embryo-sacs of other Phanerogamia and the archegonia of the Cryptogamia.

Cases are not absolutely unknown in other phanerogamic orders of the presence of more than one embryo-sac in the ovule. "*Viscum* has two or three embryo-sacs; these may all have their germinal vesicles fertilized, and the development of the embryos may go on to a certain point, until one takes the lead and the others disappear."[*] This is just what occurs in the cryptogamic spore, for though two or more of the archegonia may be impregnated, they generally all abort but one. Polyembryony may occur even in cases where there is but a single embryo-sac in the ovule, owing probably to more than one of the contained vesicles being impregnated. This is not uncommon among the Orchidaceæ, and also it is said in the genus *Citrus*, in one species of which (the orange) seeds are occasionally met with, containing more than one mature embryo.[†] That polyembryony should be of rarer occurrence among Phanerogamia than Cryptogamia admits of a very feasible explanation, from the difference in the mode of reproduction, for one result of the enclosure of the fertilizing particles of the fovilla in the pollen-tube must be to concentrate their action on a single germinal focus, and prevent the impregnation of any other, except in the rare case of two pollen-tubes entering the same ovule.

The last of the differential characters above mentioned —the association of antheridia with archegonia in the prothallium—is certainly a remarkable one. It is as if in a

---

[*] Griffith and Henfrey, Microg. Dict., p. 521.
[†] Op. Cit., p. 521.

phanerogamic plant the pollen, instead of being generated in anthers, as distinct floral organs, were formed in cells within the ovule, lying side by side with the embryo-sac containing the germinal corpuscules. It is of course needless to say that no such arrangement is met with in any plant; but, at the same time, there is observable a great variety in the mutual relations of the sexual organs. Their location in distinct individuals, by what is termed a *diœcious* arrangement, though common in animals, is rare among plants. So is also that rather closer approximation, known as *monœcious*, in which, though the stamens and pistils are in distinct flowers—or when these arise from different buds, may even be said to be in distinct phytoids— yet they always co-exist within the limits of the entire plant, or as some would term it, the *physiological individual*—*i.e.*, the product of the same original act of digenesis. But in plants it is by far most usual to have both kinds of organs within the same floral envelopes; while in what are termed *gynandrous* flowers, the anthers and the ovuliferous carpels are more or less fused together. We have only to suppose the approximation carried a little farther to have an arrangement comparable to that met with in the prothallium of a fern.

The comparative unimportance of this arrangement is farther shown by its absence in the allied cryptogamic orders of Rhizocarpeæ and Lycopodiaceæ, in which the *general* course of embryogeny so much resembles that of ferns. In Rhizocarpeæ the spore-case contains two kinds of bodies, the larger answering in so far to the fern-spores, that in germination they develope a prothallial layer containing archegonia; only, as already remarked, this is of comparatively small size, and never protrudes beyond the spore-coat. It has this farther peculiarity, more relevant to our present subject, that it developes no antherozoids, these corpuscules being the exclusive product of the smaller seed-like bodies contained in the sporocarp, which may therefore be compared to a sort of gynandrous flower,* and as the Rhizocarp fructification has this analogy to the bisexual flowers of the phanerogamia, so have some of the Lycopodiaceæ to the monœcious inflorescence, as the antheridial bodies—represented by the "small spores"—are generated in capsules quite distinct from the *oophoridia*, or those yielding the large spores, which form prothallia.

Hence the resemblance between the reproduction of such cryptogamic plants as *Selaginella* or *Pilularia*, and the phanerogamic Coniferæ, is very close in essential points, however diverse in many accidental features. In both the germinal body (the ovule or the large spore) is formed in connection with a bract or rudimentary leaf, but is quite naked, during the later stages at least of its maturation—in both there is formed in its interior, before impregnation, a peculiar cellular body, distinct from the rest of its parenchyma, (the albuminous body or the prothallium), lying just underneath the micropyle or opening in the outer coat, through which the spermatic particles gain access for fecundation—in both this new growth forms within it a certain number of cellular capsules, (the corpuscula or the archegonia), each with a germinal cell in its interior, capable of becoming an embryo on fecundation—but in both also, only one is normally so transformed, the rest aborting. In both, moreover, the spermatic cells (pollen grains or microspores) are formed in distinct organs of fructification—and in both their development is interrupted by a latent interval, represented in the former by the arrest of the growth of the pollen-tube during the winter, in the course of the two years over which the maturation of the fir-seed is extended

---

* Hence these two orders are termed *Heterosporous* by Munter, and the Ferns and Equisetaceæ *Homoiosporous*, as producing but one kind of spore, in whose cellular outgrowth are formed both archegonia and antheridia. Comptes Rendus, Dec., 1857.

—and represented in the latter by the dormant condition of the microspore for some time after being shed.* Of all the points of difference, therefore, above enumerated, the only one left outstanding in this comparison is the persistence of the ovule *in situ*, till the maturation of the embryo, as contrasted with the early shedding of the "large spore." Even the coniferous pollen-tube presents, as we have seen, changes in its contents corresponding in some degree with those which occur in the small spore of these species of Cryptogamia. We seem warranted, therefore, in arriving at the general conclusion that the spore of the higher Cryptogamia corresponds to the phanerogamic ovule, on account of the strong analogies in their progress to maturation—analogies which the distinctive characters appear quite inadequate to overbalance.†

But it is to be observed that the comparison does not extend to the so-called spores of mosses, any more than to those of the *lower* Cryptogamia or thallogenous plants. In fact, as has already been pointed out, there is no kind of affinity between the corpuscules going under this name in the two groups. As the whole process of development preliminary to the formation of archegonia is absent in mosses, the gap between them and ferns is wider—so far as the function of reproduction is concerned—than that separating the latter from the Phanerogamia. The prothallium of the fern has been shown to have its homologue within the ovule even of some flowering plants—the Coniferæ—as traceable through a distinct chain of intermediate forms, but neither to fern-spore nor to its prothallium have we any homologue whatever in the moss, where the antheridia and archegonia are directly attached to the leafy axis like the

---

* Sanderson, in Cyclopædia of Anat. and Physiol., IV. (Veg. Ovum.)
† See the abstract before given of the points both of agreement and diversity, in the notice of the Reproduction of the Coniferæ, Ch. II., § 9.

sporangia of the ferns—reminding us of the contrast between the common *Hydra* among the Polypifera, with its simple genital cysts, and the complex gonophores of other species, developing free medusoids.

The complexity of the genetic cycle of mosses is not in the processes which precede, but in those which immediately follow on impregnation, as was before shown at some length in contrasting the phenomena of alternation in these two groups. In fact there is almost as much difficulty in pointing out, in the flowering plant or fern, any homologue to the spore-case of the moss, as in demonstrating in the latter any part corresponding to an ovule or a prothallium.*

§ 10. It would apppear, therefore, that the results of the examination of the reproductive process in the Vegetable Kingdom are quite in harmony with those before drawn from a survey of the corresponding phenomena among animals; both leading to the conclusion that something which may stand for a representation of an alternation

---

* It may be thought that the suspensor of the embryo represents the seta and theca; but in the mosses and the foliaceous Hepaticæ there is, besides this, another form—the confervoid protonema—interposed before the development of the leafy axis, so that one at least of these structures must be wholly unrepresented in the higher species. In the Coniferæ there is something analogous to the multiplication of spores in the theca of the moss, for the primary "embryonal vesicle" originates four suspensors, each with its proper embryo, all, however, normally aborting but one (as in the case of the "Corpuscula" themselves), so that here, as in other plants, but a single embryo is eventually developed from the seed. On the other hand, it may be observed, that in some species of *Marchantia* among the Hepaticæ there is a stalked receptacle of the organs of fructification, which may admit of comparison with the floral organ of the Phanerogamia, or the capsule spore and prothallium of ferns. The peduncle in these cases has obviously very different homological relations from those of *Jungermannia*, for though both, when in fruit, support the sporo-cases, that of *Jungermannia* is a much later formation, being evolved from the archegonial corpuscule, while that of *Marchantia* precedes the archegonia, which are developed from the parenchymatous mass of its summit.

of forms is as generally recognizable at an advanced stage of the life of the higher organisms, in the phenomena attendant on the maturation of the reproductive organs, as we have seen it to be in the initial period of their embryonic development.

The rationale of the difference is this, that in the higher species the prevailing law of centralization prevents the structures developed for the performance of the sexual function, from ever acquiring the automatic organization and independent position that would entitle them to be recognised as distinct zooids. Such a result is confined to the lower forms of life, and indeed occurs but exceptionally even among them. Even in tribes, where alternation is the rule, there occur species in which this process of sexual maturation is so curtailed, by the suppression of structures commonly present in their congeners, that instead of separate zooids we have merely organs of reproduction of the simplest possible construction.

But the instances which have been given of transition forms, concur with the general considerations just adverted to, in suggesting the conclusion of an essential community of nature in the processes connected with the development of the structures ministering to this function, whether they put on the exceptional character of distinct zooids, or appear in the more usual form of constituent organs, in the unity of the body which produces them.

VIII.

RELATIONS OF OVA AND GEMMÆ.

§ 1. The arguments which have now been advanced for establishing a homology between the sexual zooids of some alternating species, and the more ordinary form of reproductive organs, have proceeded on the assumption that a difference between gemmæ and ova, or corresponding bodies, may in all cases be clearly ascertained, however much they may resemble each other in general appearance; and that in the latter we have a satisfactory basis from which to estimate the relative import of other structures. So much has indeed been generally admitted by naturalists, for it has been commonly held that these bodies may be recognized both by the necessity of impregnation for their farther development, and by certain peculiarities of structure.

It is allowed that these peculiarities are not always apparent at first. Thus it has been remarked—especially in orders noted for a proliferous tendency, as the Polyzoa and Polypifera—that the nascent ovum bears a close resemblance to an ordinary gemma, both in itself and in its place of origin, and other relations to the surrounding tissues. When the development, however, is complete, the most recent observations go to show that in Animals a true ovum always contains in its interior a characteristic nucleated germinal vesicle, and that in Vegetables the germ, (from which after fertilization by the spermatic particles the embryo is formed), though differing widely in different cases—as in cryptogamic and phanerogamic plants—yet admits of having its homological identity established, on structural grounds alone, by means of various transitional forms.

But the other great character (as generally considered) of bodies homologous with the ova of the higher animals— viz., their dependence on impregnation for their farther development, would seem, from recent researches, to be tenable only with considerable qualification. The facts now recorded by several independent and trustworthy observers leave no room to doubt that ova—or bodies undistinguishable from ova both in structure and relation—do occasionally undergo development independently of impregnation.

§ 2. Hence there may, under certain circumstances, be a difficulty in deciding whether the phenomena are of the nature of gemmation or of ovulation. In fact, one of the cases—that of the *Aphides*—quoted as an example of gemmation, in the course of the foregoing remarks on the different forms of alternation, has been the subject of much controversy on this very point. The question has been particularly studied by Owen, Carus, Leydig, Burnett, Huxley, Lubbock, and Leuckart, but these authors differ in some important particulars.

It is admitted that there is a general correspondence between the internal organs of generation in the oviparous and viviparous individuals; and though, as Siebold was the first to point out, there is also an important difference in the absence, in the latter, of the *colleterial glands* and *spermatheca*, yet, as these are organs ministering exclusively to the impregnation and encasing of the eggs, their deficiency does not touch on the question of essential identity or otherwise, in the process of their first formation. This is distinctly denied by Dr. W. Burnett, but, with some qualifications, is rather supported by most of the other observers.

Burnett maintains that the viviparous *Aphides* have no proper sex—that they possess no organs, external or internal, strictly corresponding to those of true females—neither ovaries nor oviducts. The germs are situated in moniliform rows, like the successive joints of confervoid plants, and are not enclosed in a special tube. These rows of germs commence each from a single germ-mass, which sprouts from the inner surface of the animal, and increases in length, and the number of its component parts, by the successive formation of new germs by a process of constriction. These rows of germs, indeed, closely resemble, in general form, the ovaries of some insects, but they are not continuous with any uterine or other female organ; they are simply attached to the inner surface of the animal, and their component germs are detached into the abdominal cavity as fast as they are developed, and thence escape outwards through a *porus genitalis*. Farther, he states that the germs have none of the structural characteristics of true ova—such as a vitellus, or a germinal vesicle and dot; on the other hand, they are at first simple collections, in oval masses, of nucleated cells, and the appearance of organization is not preceded by the phenomena of segmentation. But he admits that soon after their formation "a vitellus-looking mass is formed in connection with each."* Carus takes a somewhat similar view; Leydig again maintains that the germs of the *Aphis* do possess a germinal vesicle and macula, and that they appear to have in other respects the same internal structure as the ova. Mr. Lubbock, while admitting that he has not been able to detect these parts, appears inclined rather to ascribe this to their indistinctness in the eggs of insects generally.† Huxley and Leuckart arrive at the same conclusion, that histologically there is no difference in the germs of the oviparous and

---

* Dr. Burnett's observations are given in Ann. and Mag. of Nat. Hist., Aug., 1854, in Silliman's American Journal, Jany., 1854, and in his concluding note to his Translation of Siebold's Compar. Anatomy.

† Philos. Transact., 1857, p. 95.

viviparous *Aphides*, though they state that there are obvious differences both in the form of the ovarium, and in some collateral phenomena, such as the absence in the latter of those intermediate clusters of cells commonly termed vitelligenous, which are regarded by Dr. A. Thomson as a sort of normally abortive ova, and are compared by Dr. Carpenter to the sterile yolk segments, which form so peculiar a feature in the development of some Gasteropodous Mollusca.*

The preponderance of testimony, therefore, is decidedly in favour of the germs of the viviparous *Aphides* having essentially the structure of ova.

The same is maintained by Leuckart of the "eggs" of some Coccida, which, though not impregnated, are found to contain embryos, immediately on being laid. The reproduction of Coccida has also been examined by Mr. Lubbock, who arrives at the same conclusion—namely, that there is absolutely nothing, so far as our knowledge at present extends, to distinguish the egg formation, from that which occurs in any other Hemipterous insect.†

It appears, too, from the recent observations of Leuckart, that in the genus *Chermes*, which is also gemmiparous, and is intermediate in some degree between the two families just referred to, all the parts concerned in the formation and development of true ova are more or less represented, for the germ, which contains a very distinct vesicle, lies in an ovary of the usual type, and furnished with appendages resembling the accessory glands and spermatheca of ordinary female insects.‡

§ 3. All this must incline us, as the author just quoted

---

* Huxley, in Annals of Nat. Hist., 3d Ser., II., 215. Lubbock on the Ova of Insects, Op. Cit., III., 499; and Philosoph. Transact., Nov. 1858. Thomson, in Cyclop. of Anat. and Physiol. (Ovum).

† Philosoph. Transact., Dec., 1858.

‡ Annals of Nat. History, 3d Ser., IV., 321-411, " on the Reproduction of the Bark-lice."

remarks, to regard the germ cells and ova of the *Aphides* as morphologically identical structures; and the conclusion is strongly corroborated by the researches of Mr. Lubbock into the process of reproduction in Entomostraca, and particularly in the genus *Daphnia*.

In *Daphnia, Cypris*, and probably others of this order, a production of young has been observed to take place without impregnation—somewhat as in the case of the *Aphides*,—and, as in them, it may be repeated for many generations in succession; only the young are not born alive, but hatched from eggs. In these genera there are known to occur two kinds of egg-like bodies, one capable of spontaneous development, and the other dependent on impregnation. The eggs of the first kind are laid at frequent intervals, and in large numbers, during the summer, and very speedily develop their contents; those of the other kind are formed in sparing numbers at the close of the season, and are termed *winter eggs*, as they serve the purpose of continuing the race through the winter, being defended by their dense envelopes against the cold, which proves fatal to the parent animals, and not undergoing their evolution till the following spring. In *Daphnia*, the winter eggs are termed also *ephippial*, from their being carried for some time in an *ephippium*, or saddle-shaped mass, nearly in the same situation as the matrix for the ordinary ova, on the back of the animal, under the shell, with which they are thrown off in the process of moulting. Mr. Lubbock has shown, that while the common sort of eggs in the *Daphnia* are *agamic*—that is, capable of spontaneous evolution—the ephippial, or winter eggs, probably require impregnation. They differ also in their later development; they are then of larger size, they are enclosed in hard horny bivalve shells, and they lie in a mass of dark coloured pigment; but both kinds of eggs seem, from the observations of Huxley, and of Lubbock, to be formed originally from the same mass of parenchyma. The latter observer distinctly states that no difference can be detected in their initial development. If the common eggs are, from their producing young without impregnation, to be considered as really gemmæ, and of a different nature from the ephippial, there is nothing at first to indicate this in their structure, any more than in their substance or place of origin, for they contain an equally distinct germinal vesicle, and are formed side by side with the others, so that it is impossible to determine at first which line of development any particular germ will follow.*

In the Rotifera, there is a similar distinction of common and winter eggs, though here there is some reason to believe that it is the latter which are developed without fecundation. In this group, according to Huxley, the true ova are single cells, which have undergone a special development; the winter eggs are aggregations of cells, larger or smaller portions — sometimes, in fact, the whole — of the ovary, which become enveloped in a shell, and simulate true ova.†

§ 4. For cases affording, if possible, still more cogent arguments for the essential identity of gemmæ and ova, we may turn again to the class of Insects, where some late researches go to show that bodies, not only having all the structural characters of true ova, as in the cases quoted, but under certain circumstances acquiring, as it would seem, a capacity for impregnation, or, indeed, actually becoming impregnated, and thereupon undergoing the usual embryonic development, may at other times be evolved into normal organisms, without ever coming in contact with the spermatic particles. Phenomena of this kind have been rigorously demonstrated by Siebold, who proposes to restrict

---

* Lubbock in Philos. Transact. for Jan., 1857, and Dec., 1858; Baird's British Entomostraca (Ray Soc.), pp. 79, 87, 149.

† Quarterly Journal of Microscopic Science, vol. I., pt. 1, p. 16.

to them the term *Parthenogenesis,* originally employed in a wider sense by Professor Owen.

Siebold's remarks refer principally to some Lepidopterous insects, *(Psyche Helix, Solenobia, &c.),* and they have been confirmed by Leuckart, who has satisfied himself that the parents are true females, with ovaries of the ordinary type, and with the usual adaptations in the appendages for the impregnation of the ova, such as spermathecæ, or receptacles for the seminal fluid. The ova also have the normal structure, even in such details as the micropyle, or aperture in the shell for the entrance of spermatozoa.* But it appears to have been positively determined by both these observers that these ova may develop young without impregnation. The progeny in such cases was always female. Siebold does not appear to have determined whether these females can themselves propagate without impregnation; and there are theoretical reasons against the probability of a continuous succession of the race being maintained indefinitely, without the recurrence of impregnation from time to time. Indeed, the ascertained facts of the adaptation, both of the ova and of the maternal organs for receiving impregnation, and of the unfecundated progeny being always female, make it highly probable that in the natural course of things some are impregnated, and that in this case males are developed.

Such a conjecture is strongly confirmed by the results of Siebold's researches into the details of the reproduction of the Hive Bee, in verification of the views of Dzierzon. In this species the only perfect female and normal producer is the queen, who seems to be impregnated once for all on her marriage flight—the contents of the spermatheca sufficing for the fertilization of all subsequent ova, as required. But it would seem that only the ova destined to

---

* Microscop. Journal, Jan. 1859.

produce females are so fertilized, and that those going to form males (drones) are expelled without contact with the spermatic fluid—its access being cut off by a contraction of the sphincter of the spermatheca, which is induced in the reflex way, by a peculiar impression made on the abdomen, when inserted into such a cell as is prepared for the nurture of a drone larva.* The impregnated ova again all give origin to female larvæ, which, according to their diet and general treatment, become either fertile queen bees, or sterile workers.

The evidence on which Siebold founds his conclusion as to the non-impregnation of drone-ova is principally the following:—

1. Workers, whose organs do not admit of copulation, though generally sterile, do occasionally produce ova, but these always develope drone-larvæ. In some species fertile workers form a large proportion of the whole.†

2. It may be regarded as an ascertained fact that bees copulate in the air, and never in the hive ; and, in connection with this, it has been observed that the ova produced by queens, which, from imperfect development of their wings, are incapacitated for the marriage flight and concomitant impregnation, are also always drone.

3. The ova produced at the close of the season, even by a fully impregnated queen, are drone, the contents of the spermatheca being by this time exhausted.

4. They are found to be drone also when the spermatheca is destroyed by any injury to the abdomen.

---

* Küchenmeister, observing that it is only when the abdomen is compressed that fluid escapes from the spermatheca, suggests that the fecundation of the eggs which yield the larvæ of queens and workers may be due to the mechanical pressure of the edge of the cell on the queen's body : such pressure would not occur when the abdomen is inserted into the more roomy cells provided for the larvæ of the drones. Annals of Nat. Hist., 3d Ser., II., 490

† Leuckart, quoted in Quart. Journal of Microscop. Science, Jan., 1859, p. 104.

5. It is known that congelation destroys the vitality of the spermatozoa; and, in connection with this, it is found that though a queen, when revived after being frozen, may deposit ova, they are only such as will give origin to drone-larvæ.

6. A queen, when crossed with a male of another variety, will produce a hybrid progeny, but all the drones will be of her own variety.

7. While spermatozoa are readily discoverable in female ova, all attempts have been unsuccessful to find them in those of drones.*

The case of the hive bees goes to prove an actual convertibility of gemmæ and ova. Other cases show, with varying distinctness, a sameness of origin and of development up to a more or less advanced stage; nay, even it may be, a probability that the egg-like bodies, which ordinarily develope embryos of themselves, are occasionally impregnated; but, in the reproduction of the bee, it would seem that, up to the time when the eggs come to be laid, it is still undetermined whether they are to be impregnated or not. The Entomostraca and Rotifera lay two kinds of eggs—identical in their origin, though differing in their later development—one of which is agamic and the other not; but, in the case of the hive bee, the same ova appear to be agamic or not, according to circumstances. Out of the same brood, all ripe for fecundation, some are fecundated, some not—the selection depending on causes quite extrinsic to the organism—and both alike develop embryos, with only a corresponding difference in sex.† Phenomena such as these

---

* A true Parthenogenesis in Moths and Bees—Transl. by W. S. Dallas. An abstract of the paper is given in the Edinb. Philos. Journal for April, 1857, p. 319, by Prof. Goodsir.

† Hence, the term "agamic ovum" seems preferable to that of "pseud-ovum," proposed by Prof. Huxley, for it is a contradiction in terms to say that such a body may have already left the ovary, before it is decided whether it is to become an ovum or a pseud-ovum.

certainly compel us to qualify the commonly received statements as to the universal dependence of ova, (or bodies having the structure and relations of ova), on impregnation for their farther development. For, though we may continue to hold such to be the only bodies capable of impregnation, yet we cannot maintain, with absolute universality, the converse proposition, that every body, having the structure and relations of an ovum, requires impregnation as the starting point of its development; although we know that, in general, this is so essential, that a mature ovum, if not fertilized, very speedily degenerates, and entirely loses its vitality. In fact, the cases now referred to —confined, as will be observed, to two classes of the Articulata—are the only known exceptions to the non-development of unimpregnated ova in the Animal Kingdom.

§ 5. Similar facts, however, to these have been alleged by botanists from the time of Camerarius (1694). Spallanzani, Henschel, Bernhardi, Thuret, and many others, have adduced evidence in this direction, and more lately the question has been made the subject of special attention by Mr. John Smith of Kew, and Professor Braun of Berlin. As instances of Plants which occasionally form perfect seeds without access of pollen, Gaertner, Lehocq, and Braun cite certain dioecious species of *Zea, Mercurialis, Cannabis, Spinacia, Bryonia, Trinia, Datisca,* and *Pistacia;* and among monoecious species, the common fig, the Roman nettle, and certain varieties of melon. It appears to have been ascertained that some of these unfecundated seeds produce both male and female plants, up to the fourth and fifth generations. At the same time, it is admitted that these observations are not all rigorously demonstrative, for want of due precautions in some of them to prevent the access of pollen from extraneous sources, and to ascertain that no male flowers had appeared among the female ones—

an abnormality of not unfrequent occurrence. The most conclusive observations are those made by Smith and Braun, on *Cœlebogyne ilicifolia*, a diœcious Euphorbiaceous plant from Australia, of which the male has never been introduced into this country, and in which there is no tendency to the occasional development of stamens. In this plant the maturation of seeds without pollen has now been a subject of notice for upwards of twenty-seven years, and it appears, from the observations of Deecke and Radlkofer, that the structure, both of the ovule and of the mature seed, are perfectly normal, and that the embryo follows the usual course of development, with the exception of the absence of the pollen tube. Radlkofer thus sums up the arguments in proof of a "true parthenogenesis" in this plant :—

1. The ascertained absence of all male flowers of the same species.

2. The absence of all indications of hybridization by males of allied species.

3. The development of the embryo, without any trace of a pollen tube in the ovule.

4. The stigma not withering, as it always does after impregnation. This mark indicates parthenogenesis in some of the other plants experimented on, especially *Cannabis* and *Mercurialis*. The seeds thus raised have never yet given origin to any but female plants. Braun thinks that similar phenomena can be shown to occur in some diœcious species of *Chara*.*

§ 6. Intermediate between cases such as these, of the

---

* Radlkofer, in Annals of Natural History, 2d Ser., XX., 216.
Balfour, in Edinburgh Philosoph. Journal for June, 1858, vol. VIII., 159.
Lubbock, in Philos. Transact. for 1857, page 97.
Owen's Address to British Association at Leeds, 1858, p. 26.
For Regel's arguments against such Parthenogenesis, see Annals Nat. Hist., 3d Ser., III., 104.

development of the ovule without impregnation into a fertile seed, and ordinary gemmation, may probably be ranged what are called *viviparous* flowers. In these buds are formed within organs having the general character of floral envelopes, where they occupy commonly the position of the ovary in a normal flower. This peculiarity sometimes constitutes a specific or constant character, as in the *Polygonum viviparum*, in which, while the upper bracts of the floral spike bear the sexual flowers, the lower ones bear gemmæ, ripening without impregnation into deciduous bulbils, and acting the part of phytoids in the propagation of the plant. More commonly, perhaps, the viviparous condition is induced by the force of circumstances, as a compensating provision under a state of matters liable to interfere with the maturation of the essential parts of fructification—the stamens and pistils. Thus certain grasses which have regular flowers in genial habitats, are found to present the viviparous inflorescence in alpine situations; and cases are even met with in which the same individual plant forms at one time gemmæ, and at another time ovules, within its floral envelopes, according to variations in climate or other external circumstances. If we may assume in such cases that the viviparous condition is in compensation for the inability of the plant to form perfect stamens, there would arise a certain probability of the actual transformation of the rudimentary ovule into a gemma, capable of vegetating into a new plant, apart from any fertilizing action of the pollen.

For farther arguments on the essential identity of ovules and gemmæ, reference may be made to works on Vegetable Morphology, in which this view is almost universally adopted. So much at least appears to be shown by the occurrence of viviparous flowers, that the same focus of vital action may be capable of assuming the form of a gemma, to propagate the plant by its unaided plastic powers—or of becoming an ovule, whose development is dependent on impregnation. In such cases it would seem that, though in the normal course of development preference is given to the seed as being the surest means of perpetuating the species in all its integrity, and as possessing a more enduring vitality when separated from the parent plant, yet under peculiar circumstances, the growing point may be transformed into a gemma whose early evolution makes less demand on the parental vitality, than the impregnation of the ovule and the formation of the embryo.

§ 7. Though it is, no doubt, quite exceptional to meet with cases such as have now been cited, in which a body, with the structure and relations of an ovum, either passes into a true gemma, or acquires the power of self-development characteristic of gemmæ, traces of an incipient tendency in this direction are not wanting in many cases, in which impregnation is absolutely essential to full development, and may possibly turn out to be a more frequent characteristic of the sexual elements of both kinds than is generally supposed. Thus though in ordinary cases unimpregnated ova do certainly soon lose their vitality, yet they have been seen to make in the interval some abortive attempts, as it were, in the way of development. Thus it has been observed, that, though the unimpregnated spores of *Fucus* never produce fronds, yet they may put forth irregular prolongations as if about to germinate.* And so in the unimpregnated ova of animals, the preliminary phenomena of segmentation have been observed by Loven and Schultze in *Campanularia*,† by Quatrefage in *Hermella* and *Unio*,‡ and by Vogt in *Firola*.§

* Carpenter's Comparat. Physiol., 4th Ed., p. 493. (Thuret.)
† Quar. Jour. Micr. Sc., III., 65 (from Muller's Archives).
‡ Rambles of a Naturalist, II., 244, and Comptes Rendus, July 23, 1849.
§ Siebold on Parthenogenesis, 106.

In the Batrachian ovum, on the other hand, Mr. Newport's researches led him to conclude that segmentation never takes place without impregnation, but only certain preliminary changes of a less obvious kind. Even in impregnated ova, when, from the excessive dilution or the momentary contact of the spermatic fluid, this process is inefficiently performed, the segmentation stops at its incipient stages.* But this result is only what we should anticipate in the ovum of the Vertebrata, in accordance with the general abeyance of gemmation in that division of the Animal Kingdom.

These facts seem to indicate rather an evanescence, or a feebleness of vitality, than an absolute want of developmental nisus in the unimpregnated ovum, as the cause of its general abortion.

But that the spermatic element is something more than a mere stimulus is shown conclusively by the phenomena of hybridization, and the general fact of the transmission of the paternal peculiarities to the offspring. There are not wanting indications also of a certain *intrinsic* capacity of development in the spermatic element, at least in plants. Setting aside the power of germination, asserted by some for the spermatia of Fungi,† and antheridia of Mosses,‡ the continued growth of the extremity of the pollen-tube in impregnation, even after the decay of the granule itself, and, according to Schacht, its occasional branching,§ are well attested facts which look in this direction. It will probably be found also, as observations are multiplied, that the zoospores or motile gemmæ of Algæ merge by as gradual a

* Phil. Transact. (1851), p. 169.
† Berkeley's Introduction to Cryptogamic Botany, 247, 259, note.
‡ Lindley's Natural System of Botany, p. 407.
§ Schacht on the Microscope (Currey), p. 144. Reissek maintains the occasional development of pollen grains into fungoid vegetations. Quoted by Sanderson in Cyclopæd. Anat. and Physiol. "Vegetable Ovum," p. 255, note.

transition into the phytozoa or ciliated spermatic particles, as the fixed gemmæ of some species of articulated animals do into true ova.

It may be observed also that although the vitality of the spermatozoa is soon lost on exposure, it may be protracted indefinitely under more favourable conditions, as by transference to the female organs. One charge of the spermatheca will enable a queen-bee to lay impregnated eggs for several years, during the whole of which time the spermatozoa continue in a state of activity. Siebold quotes Dzierzon for the statement that a queen may acquire the power of laying fertilized eggs for five years by a single normally executed copulation.* The evolution of the spermatozoa, indeed, is only in progress in some cases at the ejaculation of the seminal fluid, and is thus perfected in the normal course of things after their removal from the male organs. Thus in Spiders these particles are not matured till after the fluid is transferred to the cavities of the palpi, and in the higher Crustacea not till its introduction into the passages of the female.†

§ 8. May not the true theory be, that each of the sexual elements is essentially a gemma, endowed with a high degree of vitality, which confers upon it a capacity of development, but which is, at the same time, of a self-exhausting nature, unless tempered by combination with its prearranged complement of the other sex—the fusion of the two elements being a sort of physiological illustration of the connubial relation of the sexes, "ordained for the mutual society, help, and comfort, that the one ought to have of the other."

---

* True Parthenogenesis in Moths and Bees, p. 61, note. It is the same also with several Arachnida, according to Blanchard. Comptes Rendus (April 9, 1860), p. 127.

† Wagner and Leuckart, Cyclop. of Anat. and Physiology, article, "Semen." H. Goodsir, Anatomical and Physiol. Observations, p. 39.

Such a view harmonizes with other results obtained by Mr. Newport.* Thus he has observed that a deficiency affecting one of the elements may be supplemented by an increased action of the other, for he found that when the spermatozoa were applied in a very concentrated form, even immature ova became segmented and produced embryos. Again he found that the procreative force both of the ova and the spermatozoa was augmented by increase of heat, but the duration of the force was lessened. Now, in the primordial germs, that physiological law may well be supposed to have full play, which appears even in the adult state—though there masked by the continual renewal of the system, which it is the very object of its organization to effect—namely, that the endurance of vital action is in the inverse ratio of its intensity.

On the whole, after a careful survey of the part taken by ova and gemmæ respectively in the propagation of organic beings, I think we can hardly avoid adopting Professor Owen's conclusion, that there is no *essential* difference between the two, and that the one may pass into the other by insensible gradations. It cannot at least be any longer maintained as an invariable law—though it is strongly contended for by some—that there is such an incompatibility between these two modes of propagation, that in proportion as any part of the parenchyma of the parent body is engaged in the one course, it is in so far disabled for the other. Such a statement, if admitted at all, must be held open to many exceptions. We may assume perhaps, that, up to a certain point, the development of the new focus of vital action may go on all the same for a gemma or an ovum ; but that towards the period of maturation the changes which take place in the latter, to fit it for impregnation, cause such a tension, as it were, of its vitality, as is incom-

* Philos. Transac. (1851), pp. 206-209.

patible with its continuance, in the majority of cases, unless re-invigorated by the access of the spermatic element.

§ 9. This doctrine of an essential community of nature between ova and gemmæ has an obvious bearing on our views regarding the modes of propagation in which they are respectively concerned. If the nascent bodies approximate so much in their real nature as to pass into each other, the processes of development can hardly be so widely different in the two cases as is held by some.

The prevailing view in this direction seems to be that gemmation perpetuates the individual rather than the species, the successive zooids or phytoids preserving, more completely than the progeny of embryonic origin, the characters of the parent stock ; and it has been thought, too, that there is a tendency for the plastic power to wear out in process of time, so that a recurrence of sexual generation at intervals is necessary to preserve the pristine vigour of the species. The more highly organized the species, the more dependent it is supposed to be on the frequent recurrence of sexual reproduction in the genetic cycle. In the higher animals we meet with no obvious phenomena at all of the nature of gemmation, while in the lowest there may be a very prolonged pullulation of gemmæ, the sexual act recurring only at distant intervals. Still it is held that even in these it must recur from time to time, to give a fresh start, as it were, to the organizing process. It is true there are species in which we have no positive knowledge of any sexual act occurring at all—such is the case with most of the Protozoa. There are species, too, in which no males have yet been detected—so it is with some Insects and Entomostraca.* But such negative evidence is not held of sufficient weight to counterbalance the argument

---

* Certain species of *Cynips, Apus, Limnadia,* and *Polyphemus* are in this case. Siebold's Parthenogenesis, 105.

from analogy, being put down to the account of imperfect observation.

The great difficulty, indeed, in the way of admitting a total and continued absence of the sexual act in any species is its apparent contrariety to the general analogy of the organic creation. The constancy of the presence of both sexes in almost every species, and the remarkable adaptations which are provided in particular cases to ensure impregnation, certainly suggest the idea that the act is one of essential importance to the perpetuation of race. The act may take place with less frequency in certain cases—indeed it is well known that in some species the occurrence of males is rare and intermittent—but still it is one which we have difficulty in conceiving ever to be wholly dispensed with.*

Now the cases of Parthenogenesis, above referred to, do not appear materially to affect this conclusion. They still leave the question open of the necessity of an act of sexual reproduction recurring from time to time in the continued propagation of a species. In the case of the bees it is clear that the act of monogenesis must alternate with one of a sexual kind, as the unimpregnated ova produce only male insects. In all instances, indeed, yet observed, agamic broods have been always of one sex in the same species, a circumstance which is in favour of the occasional recurrence of the sexual act, as the probable determining cause of the development of the other sex.†

---

* Professor Braun of Berlin, in the Transactions of the Royal Prussian Academy (Annals of Nat. Hist., 2d Ser., XVI., 234), mentions as a somewhat parallel case to the intermittent recurrence of male Entomostraca, the casual appearance of staminiferous flowers in the weeping willow (*Salix Babylonica*), which has always been propagated by cuttings derived originally from one female plant, and which ordinarily bears only pistilliferous catkins.

† This may hold of normal agamic broods, but Siebold found that when the unimpregnated eggs of the silkworm-moth proved fertile, the resulting

In regard to the length of time for which agamic propagation may run on, our observations are as yet very imperfect. Lacordaire has followed the parthenogenesis for three generations in *Liparis dispar*,* and Davis to the same extent in the Egger Moth; all the progeny in both cases appear to have been female.†

Many plants, it is known, may be propagated by buds or cuttings for a long series of years, without ever flowering or seeding; in some species of shrubs and trees, which have long been cultivated in our gardens, only one of the sexes is known to exist in this country—the *Aucuba Japonica* and the Weeping Willow are instances in point.‡

§ 10. In conclusion, a few remarks may be made on the bearing of these phenomena of "true parthenogenesis" on the theory of alternation of generations. Startling as are many of the facts recorded by Siebold and others, and much as they have tended to modify the opinions formerly held on the relationship of ova and gemmæ, they do not seem to be at all opposed to the parallel which has been traced in the different forms of alternation, and the phenomena of embryogeny and sexual maturation in the higher species. On the contrary, they appear to supply the very links wanting to the completeness of the argument.

---

caterpillars developed both male and female moths. The inferiority, however, of the organizing force in the unimpregnated eggs was clearly shown by the small proportion in which an embryo was formed, and by its often aborting even when formed; this, of course, must militate against the perpetuation of the species independently of the sexual act. "True Parthenogenesis," p. 100, &c.

* Lubbock, in Philosoph. Transact. (1857), p. 96.
† Siebold on Parthenogenesis, p. 99.
‡ Among the lower plants instances of this kind are even more striking. The Gulfweed, which accumulates in such quantities as to acquire for its oceanic habitats the name of "floating meadows," multiplies solely by the detachment of offshoots, and has never been found in fructification (R. Brown, in Annals of Nat. Hist., 2d Ser., VII., 327). So it is also with many mosses, as noticed in Chap. II.

The formation of ova has been here considered as in its initial stage a modification of gemmation, in connection with the maturation of the reproductive organs—these forming the parent stock, as it were, from which the gemmæ in question may be said to sprout. But, had we no examples of ova undergoing development in virtue of their own plastic energy, we should miss in them one of the most essential characters of gemmæ. As the matter stands, however, nature has provided us with a complete series of examples, establishing a well-marked transition from simple gemmæ, undergoing evolution in the ordinary way, to perfectly normal ova, absolutely dependent on fertilization by the spermatic element.

The view of the genetic cycle, which has been put forward, is based on the existence of three successive stages in the life-history of all organic beings, of which the first is preliminary to the full development of the typical organization of the species in the second, and the last has for its great object the perpetuation of the race, by elaborating and providing for the fecundation of ova endowed with a special retentiveness of vitality and aptitude for development. In one point of view, these three stages in the life of each individual may rather be regarded as a succession of so many distinct individuals, each derived from the preceding by an act of gemmation :—using this word in its widest sense, to imply simply the establishment of a new focus of constructive and other vital action, in an extension of the substance of the parent body. The new growth may be single or multiple ; and it may separate from the parent as a distinct organism, or it may in some way absorb it or take its place. From variations in these points we may have the alternation and multiplication of diverse forms, or we may have merely successive phases of development in a single individual. But, in all cases, the derivation of each stage from the preceding may be held to be by some process of the nature of gemmation; and the close affinity, if not identity, which has now been shown to exist between the processes of ovulation and gemmation is of especial interest in this respect, that it indicates a parallelism between the origination of the protomorphic stage, from the gamomorphic structure of the *preceding* cycle, by the process of generation, and the derivation of each of the other phases from that which immediately precedes it in the *same* cycle. Just as the typical form is budded off from the germinal, and as the gamomorphic form itself is budded off from the orthomorphic, so is the germinal stage of the next generation budded off in the form of an ovum from the female ovary.

That there should be certain specialities in this particular gemmation is only what might be anticipated, when we consider that the great function of this stage of existence is the perpetuation of the race. Whenever the continuity of the vital processes is broken in upon by a period of latency, we find almost universally that this break follows close upon ovulation, which is the characteristic gemmation of the gamomorphic stage. Such a state of latent vitality is of essential importance in the great majority both of animals and vegetables, serving as it does to bridge over seasons and circumstances which would otherwise prove fatal, and affording, in the form assumed—that of eggs or seeds—peculiar facilities for the dispersion of the species. An intermediate process of pullulation may occasionally occur in the gamomorphic stage, but eventually eggs or seeds are formed in almost every case.* An egg is itself essentially a modified gemma, derived from a gamomorphic structure; a seed is the persistent wall of such a gemma which has already developed other structures in its interior; but both, being destined for the continuation of the species, under

* The Ferns and their allies constitute, perhaps, the only case in which the reproductive organs remain permanently in adhesion to the typical structures developed by their instrumentality.

special conditions, have naturally certain peculiarities in their mode of origin different from those of ordinary buds.

Among such peculiarities is their development, in connection with the organs of sex. If there be any truth in the opinion before referred to of the necessary recurrence of the sexual act from time to time in every species, this of all seems the fittest season, when bodies of such importance are to be formed. Such, at least, is the case; whether the explanation be in point or not, it remains a fact, that the proper gemmæ of the gamomorphic stage have always a sexual character. Ordinarily, impregnation is a necessary condition of their ulterior progress; the transformation by which the gemmæ are converted into spermatic and germinal elements being apparently incompatible, in most cases, with the farther development of either singly. But instances, as we have seen, do occur, in which, either in compensation for the failure of the spermatic element, or for other reasons, the co-related gemma is fitted to retain the power of self-development. Still, however, it acquires somewhat of the sexual character, and may, indeed, in some cases, take on all the structural peculiarities of the ovum, and even its capacity, in certain contingencies, for impregnation. In every case, there probably remains at least such a speciality of character as clearly to distinguish these gemmæ from those developed in any other stage of the genetic cycle.*

* An exception possibly occurs, in a few species of Polypifera, in the development of the gamomorphic structures themselves (the medusoids) from gemmæ having a general resemblance to ova—but the facts of the case are not clear of doubt. See before, in the note at page 129.

IX.

SUMMARY OF CONCLUSIONS.

§ 1. Having discussed, in the preceding chapters, the distinctions observable among cases of alternation of Generations, and the phenomena which may be thought in some way to represent them in the higher species, and having stated the general results of later observations in reference to the relations of ova and gemmæ, it may be advisable to give here a short summary of the conclusions arrived at, as affording that explanation of the facts which appears, on the whole, the most probable, if not in all cases positively ascertained. They may be stated as follows :—

I.—The function of reproduction is performed in two ways—by gemmation (monogenesis), and by sexual union (digenesis.) In the former, a portion of the body of the parent becomes the seat of a certain independent vitality, and may eventually be detached and transformed, by a process of self-development, into a distinct organism (zooid or phytoid). In the latter, the germ of the new being is the result of fecundation—that is, of the fusion of two highly organized portions (spermatic and germinal) of the same or of kindred organisms, which stand in the relation of parents.

II.—There is reason to believe that the same portion of the parenchyma of the parent may be organised in either of these ways; but, in general, in proportion as it is engaged in the one course it is disabled from the other; though a few cases are known in which a part already organized as an ovum may yet be developed as a gemma, that is, without impregnation.

III.—Reproduction, by fecundated germs, appears to be of a higher kind than that by gemmæ, and has been regarded as the special agency for the propagation of the species, the other serving rather for the extension of the individual.*

IV.—There may be reason to believe that the recurrence of sexual generation, at least occasionally, is necessary for the long-continued propagation of every species ; but, in many of the lower forms of life, it occurs but rarely, and at distant intervals. In the higher, again, it is the prevailing mode, and the counter process of gemmation is minimized, and its manifestations often very obscure.

V. In those in the lower species, in which both modes of propagation are well marked features, we find that they have a tendency to succeed each other in a regular order, to which the term of "alternation of generation" (metagenesis) has been applied.

VI.—The phenomena of alternation of generations present certain remarkable diversities, dependent in great measure on the period of the life-history of the species at which a distinct process of gemmation is interpolated.

VII.—In some cases the interpolation takes place in what has been termed here the protomorphic stage—that is, in the course of development prior to the appearance of the typical character of the species (as in Trematoda and Mosses.)

VIII.—More frequently, perhaps, a process of gemmation is interpolated in the later or gamomorphic stage—that is, in connection with the development of the reproductive

---

* This idea, based partly on the greater resemblance to the parent, in most cases, of the progeny developed by the budding process, has the support of many great authorities. "*Gemmæ individium continuant, cum semina speciem propagent,*" is an aphorism of Link (Elem. Philos. Bot., p. 208) ; that of Linnæus is different—"*Gemmæ totidem herbæ.*" Braun's Rejuvenescence, pp. 19, 24.

organs, after the typical form of the species has been acquired (as in the Polypifera and Ferns.)

IX.—In the orthomorphic or intermediate period, the gemmæ generally remain in adhesion, to form a compound organism (as in zoophytes and most plants), but a few cases are known in which they give rise to a well-marked alternation of distinct forms (as in the *Aphides.)*

X.—It is but rarely that alternation occurs in more than one stage of the development of the same species. Even in the larger groups or orders characterized by such phenomena, their marked occurrence is mostly confined to one of the life periods above specified. Among the Cestoid worms, however, we find distinct alternation, both in the protomorphic and gamomorphic stages; and in the Polyzoa, and some Annelida, less obvious processes of the same kind appear to occur in all the stages.

XI.—The gemmation, which occurs in any stage, may occasionally be lengthened out by the successive pullulation of a series of zooids or phytoids of the same general character, previous to the supervention of the characteristic phenomena of the next succeeding stage. The number of interpolated links is fixed in some species, and variable in others. Such pullulation may occur at any stage, but is most common in the orthomorphic, giving rise in it to the compound ramified structures characteristic of plants and zoophytes.

XII.—In the embryogeny of the higher species, a certain parallel may be traced to the first or protomorphic form of alternation, in the implantation of the primitive trace of the typical organization, on the cellular mass into which the contents of the ovum are resolved, by the process of cleavage following immediately on impregnation. This view is founded on the following considerations:—

1. The duplication, in whole or part, of the embryonic axis, as an occasional abnormality, even in the higher species, resulting in the formation of a double monster.

2. The normal formation of a double embryo from the ovum, in the case of the Polyzoa.

3. The variable character of the gemmation in the Cystic Entozoa, commonly solitary, as in *Cysticercus;* but multiple in some, as in *Cœnurus* and *Echinococcus.*

4. The co-existence among the Echinodermata of cases of ordinary embryogeny, with others, in which the embryo, though still solitary, has a character of distinctness from the primary matrix, or so-called larva, which has led to its being regarded as an independent zooid.

XIII.—To the gamomorphic form of alternation, a similar parallel may be traced in the development of the reproductive organs of the higher species. Such a correspondence is suggested in particular by the following considerations:—

1. The periodicity and lateness of development of the organs of reproduction in most species; and their greater or less independence of the rest of the system in some cases, as in the Polyzoa.

2. The transition, among nearly allied species of certain families, from cases in which the reproductive organs are integral parts of the system, to others in which they occur in detached zooids having the character of distinct and well-organized animals.

3. The accidental or non-essential nature of the characters of detachment and organization, which principally distinguish these zooids from common organs of reproduction, as is illustrated in the orders just referred to ; and as is indicated, further, in respect of the latter point, by the contrast of such forms as the rudimentary males of the Cirripedia and Rotifera, which, though truly distinct animals, have their structural development limited almost exclusively to the generative organs.

4. The correspondence traceable between the *embryosacs* of the *ovules* in phanerogamic plants, and the *archegonia* in the *prothallia* of ferns and allied Cryptogamia.

§ 2. If it is desired to have a formulated expression of the sequence of phenomena in the genetic cycle, the following may be taken as a sort of medium statement:—

I.—PROTOMORPHIC STAGE.

Development of the fecundated Germ into a rudimentary asexual organism, serving as the matrix of the typical form.
Gemmation, or formation in connection with this matrix of one or more growths of a new and different kind.—*(Scolex, Redia.)*

II.—ORTHOMORPHIC STAGE.

Development of such a growth into an organism of the typical form of the species.
Gemmation, or formation in the typical organism of new points of growth, for the development of reproductive organs.

III.—GAMOMORPHIC STAGE.

Development of the new growth into a more or less complex structure, serving as a receptacle of true sexual organs *(Medusoid, Prothallium, &c.)*
Gemmation, or formation in such structures of spermatic and germinal bodies, the latter, on fecundation by the former, constituting the starting point of a new cycle.

§ 3. In applying this formula to the case of the higher species, the protomorphic matrix must be held as represented by the cellular mass, resulting from the cleavage of the germ on impregnation, and serving as the basis of the true embryonic structures. And so, in the gamomorphic stage, the representatives of the reproductive receptacles must be sought for in certain organs of the body, often not at all conspicuous, the *spermaries* and *ovaries*, or, perhaps, rather the minute spermatic cysts and ovisacs which are developed in their interior at the appropriate season.

Hence, the formula, modified to the case of the higher animals, would stand as follows :—

I.—PROTOMORPHIC STAGE.

Development of the fecundated germ into a peculiar cellular mass, by a process of cleavage—*("Mulberry body.")*

Gemmation, or implantation of the primitive trace of embryonic organization in the *area pellucida* of this germinal mass.

II.—ORTHOMORPHIC STAGE.

Development of this primitive trace of organization, through the successive stages of embryonic growth, into the typical conformation of the species.

Gemmation, or formation, in the typical form, of a matrix, which is subsequently to originate reproductive organs of greater or less complexity—*(Wolffian Body ?)*

III.—GAMOMORPHIC STAGE.

Development of the true organs of reproduction—spermsacs and ovisacs.

Gemmation or formation, in their interior, of vesicles of evolution containing spermatozoa, and of ova with germinal vesicles ; by the joint action of which, in impregnation, a basis is formed for a new cycle of changes of the same kind.

§ 4. On the other hand, a ffarther expansion must be given to the formula, to adapt it to cases of repeated pullulation. For this purpose a middle term must be inserted under each head ; for, as has been shown, a process of pullulation may be interpolated in any of its three stages. Thus modified, the formula will stand as follows :—

I.—PROTOMORPHIC STAGE.

Development of the fecundated germ into a rudimentary form.
> *Pullulation*, or production of a continued series of gemmæ, which are either detached, or remain in connection with the stock as a compound organism—(*Protonema.*)
> *Gemmation*, or formation, in the concluding links of the series, of new points of growth.

II.—ORTHOMORPHIC STAGE.

Development of such gemmæ into organisms of the typical form of the species.
> *Pullulation* to a variable extent in this form—(a *Polypidom* or *Vegetation.*)
> *Gemmation*, in the concluding pullulations, of fresh points of growth, for the evolution of sexual organs.

III.—GAMOMORPHIC STAGE.

Development of the latter growths into sexual structures of greater or less complexity.

Pullulation of similar derivative structures to a variable extent.

Gemmation ultimately of the proper reproductive organs—sperm sacs and germ sacs—containing their characteristic corpuscules, for the purpose of fecundation.

§ 5. This formula, and that at p. 216, as constructed to include the different forms of alternation of generations above referred to, of course has not, in general, all its details obviously represented in any one species. The alternation comes out prominently in one stage, while the corresponding changes are more or less suppressed or curtailed in the others. It is, perhaps, of most frequent occurrence in the gamomorphic stage, next in the protomorphic, and least so in the orthomorphic; while there appear to be but few cases, as already mentioned, in which it is distinctly met with in all the three stages.

In one of the tables given in the Appendix, the orders best known, for examples of this kind, are arranged under the heads of the life periods in which the more obvious phenomena of alternation occur. In those that follow, an attempt has been made to arrange, under the same heads, some of the most striking phenomena in the genetic cycle, both of alternating and non-alternating species of plants and animals.

It may be mentioned, in conclusion, that however great the effect of these processes of gemmation and pullulation may be in modifying the external form of those species in which they prevail extensively, little or no relation can be traced between the peculiarities of the genetic cycle and the affinities of the species, as recognized by systematic naturalists. We have seen that totally different laws of gemmation prevail in nearly allied orders, as in Cestoid and Trematode worms among animals, and in mosses and ferns among plants ; and we find that, even in the same families —and these highly natural ones—free zooids occur in some species and not in others, closely allied ; of this, many illustrations have been given in the course of the foregoing remarks. The occurrence of pullulation is even more irregular, not serving always even as a generic character ; thus, we have arborescent, herbaceous, and even annual species in the same genus, as in *Solanum, Veronica,* and *Hibiscus.*

X.

CASES SIMULATING ALTERNATION OF GENERATIONS.

§ 1. In connection with the subject of alternation of generations, a short reference may be made here to various cases, which, without being really of the same nature, have some apparent resemblance to it, or have for some reason been confounded with it.

Most properly the expression is used to imply a certain regularity of recurrence of sexless and sexual forms in the succession of progeny, and a certain diversity in their general character. It cannot be regarded as a proper case of alternation of generations, when the interchange of the two modes of propagation is casual and not a fixed character of the species—as may be the case, for instance, in plants with bulbiferous stems or viviparous flowers, which propagate by gemmæ in seasons when they cannot mature their seeds.

§ 2. Nor again does an interchange of different modes of budding belong to the class of cases we have been considering, for this would be an alternation of *gemmations* rather than of generations. Instances of such a kind occur in the alternation of attached and deciduous buds in plants. Thus a grass raised from one of the bulbils thrown off by a deciduous inflorescence multiplies its stems into a large tussock by a succession of leaf-shoots which vegetate in connection with the primary axis, and may all bear gemmiparous flowers, like that which furnished the original deciduous gemma.

In many of the lower algæ we find a similar alternation of free and attached gemmæ, in the form of zoospores, and of cells cohering to form confervoid filaments. And much in the same way we have in the Polyzoa, besides the attached Polype-bud, a peculiar form of deciduous gemma—the *statoblast*.

In the cases cited, the two forms of gemmæ which thus take part in the propagation of the species differ in the matter of attachment to the parent stock as well as in their own character, but cases occur in which both are fixed, as well as others in which both are free. Thus, in some species of *Sertularia* we may have a frond of seaweed covered with a miniature forest of small polypidoms, each with its own polype-buds, but all at the same time organically connected by stolons proceeding from radical gemmæ, which creep in a network over the surface of the frond. Two forms of free gemmæ again appear to co-exist in some Algæ, for there is reason to believe that the statospore, which is often associated with the zoospore, and is intended for the reproduction of the species after a considerable lapse of time, is not in every case the result of impregnation, but is sometimes rather a true gemma, modified to meet the peculiarities of the case.

§ 3. As an alternation of different forms may thus occur in a succession of gemmations, without the intervention of any sexual process, so it has been thought that a periodic recurrence of certain peculiarities may be observed in the course of generations, even where the only mode of propagation is by the formation and impregnation of ova. In our own species something of this kind has been recognized under the term *atavism*, by which is meant the reappearance of particular features, tempers, morbid diatheses, &c., in the progeny of the second generation, passing over the immediate offspring. Facts of this kind are of frequent observation—a man's resemblance to his grandfather is a common subject of remark, as well as the tendency of certain hereditary diseases, as gout, to reappear in the grandson, rather than in the immediate issue.

§ 4. By some authors the phenomena of *Metamorphosis*, as they occur in Insects, Crustaceans, Batrachians, &c., have been confounded with those of alternation. Such metamorphosis is described by Dr. A. Thomson as a breach of continuity in the progress of embryonic development, "marked by some change in the mode of life, or some difference in the structure of the individual"*—a definition which might perhaps be stretched to include cases of alternation. In almost all cases of metamorphosis there is a separation of what was originally a single body into more parts than one, inasmuch as there is always something cast off—either, as in Insects and Crustaceans there is *ecdysis* or casting of the skin (what Huxley terms concentric fission†)—or, as in the Ascidians and some Batrachians, the casting off of a member, as the tail. The difference in so far is perhaps one rather of accident than of essence, and seems to consist in this, that in the metamorphosis none of the parts but one is capable of continuing its existence, so that, disregarding all the others, we at once identify the living segment with the animal which has been the subject of the metamorphosis; whereas in well-marked cases of alternation, the vitality being more equally divided, we have in the swarm of resulting zooids, so many candidates for identity with the pre-existing structure, that we involuntarily cancel their claims as destructive of each other's validity, and regard them all alike as the *progeny* of the pre-existing structure, and not as a *continuation* of its very self. That there is not merely this community in theory, but that in actual nature the two cases do tend to merge into each other has already been pointed out in speaking of

* Cycl. Anat. and Phys. Sup. [131] (ovum).
† Lect. at Royal Inst. Ann. Nat. Hist., 2d Ser., IX., 505.

the Echinodermata. Whenever, as in this class, but one new zooid results from the alternation, it is obvious that the great difficulty of identifying it with the prior structure is removed, and in some of the species of Echinodermata the process has in fact much more the character of a metamorphosis than of an "alternation." In the development, for instance, of the *Echinus* from its "Pluteus" there is less rejection of the pre-existing structure than occurs in the exuviation of the skin of the Caterpillar, the casting off of the tail of the Tadpole, or even in the removal of the placenta of the Mammalian fœtus; but in the Starfish—a close ally of the *Echinus*—though we still have but a single new zooid, our sense of continued identity is obscured, if not destroyed, by the structure in which it originated, retaining its own vitality, and appearing as a distinct animal, side by side with that which has issued from it.

After all, it is not metamorphosis which is proved in this way to be akin to alternation, but only that casting off of old structures simultaneously with the formation of new, which, though a common accompaniment of metamorphosis, is far from being its prominent characteristic; for the developmental changes in which the process essentially consists tend much more to the fusion of parts previously distinct, than to the breaking up of prior continuity of structure. This is particularly remarkable in the embryogeny of those species, in which metamorphosis is best marked, as Insects and some of the Crustaceans. On the other hand, it is not in metamorphosis merely, or as a progressive step towards higher organization that we have a casting off of certain parts of the body; exuviation at least of the tegumentary tissues is a phenomenon, which, continuously or intermittently, and in a greater or less degree, goes on during the whole lifetime of the individual in almost all species, and is in fact absolutely the same in kind with that constant molecular renewal of the body, in which the very essence of organic life consists. One does not see why, for instance, the exuviation of sheets of cuticle after scarlet fever is not as much entitled to be called "a concentric fission" as the ecdysis of an insect in the course of its metamorphosis.

§ 5. Another class of cases which in some degree simulate those of alternation consist in the formation of *spermatophores* and *nidimentary envelopes*. These structures, which are sometime elaborately organized, have been met with in Cephalopoda, Gasteropoda, Insecta, Crustacea, Annelida, Planaria, and perhaps other Invertebrata. In the Cephalopoda the spermatic corpuscules, when generated in the usual way, become enclosed in peculiar cases or receptacles, which in the case of the *Lolligo* have been long known to naturalists under the name of "filaments of Needham." They are of a very complex structure, but their motile powers, chiefly shown in the ejection of their contents, may probably be accounted for on physical principles.* In the *Argonauta*, in which the spermatophore is a simple sac, its active functions are discharged by the peculiarly modified tentacle *(Hectocotylus)* in the cavity of which it is lodged previously to the member being thrown off from the body of the Cephalopod.† Spermatophores of

---

* In *Nautilus* the spermatophores after ejaculation appear to be coiled up within the funnel into peculiar laminated vesicles. When uncoiled they appear as attenuated filaments about half a line in diameter, but of great length—*e.g.*, 27 to 34 centimetres in *Nautilus*, 3 feet in *Octopus Carena* (Leuckart), 8 centimetres in *Octopus vulgaris* (M. Edwards). Hoeven, in Ann. Nat. Hist., 2d Ser., XIX., 72.

† Annals of Nat. History, 2d Ser., IX., p. 492. Ann. des Sciences Nat., i. XVI., No. 3. From the observations of Steenstrup it seems probable that in all male Cephalopoda one of the eight shorter arms is different one side of the body from its fellow on the opposite side, and is developed for a certain extent of its length in a manner more or less analogous to that of the Hectocotylus-arm of the *Argonauta*. The particular arm and the side of the body affected vary in different species. Annals of Nat. Hist., 2d Ser., XX., 81.

a simpler structure are met with in some Insects, Crustaceans, and other Invertebrate animals.*

Much in the same way the eggs of some Insects, Annelidans, and Molluscs,† after acquiring the usual character of impregnated ova, become enclosed in peculiar envelopes; and some of these structures also are said, though on rather doubtful evidence, to have certain motile powers.

In the simple fact of the enclosure of the reproductive bodies in appropriate receptacles there is nothing very peculiar; it is merely, so to speak, the application to masses of spermatozoa and ova, of what we are familiar with in the case of single ova, in the formation of the shell of the bird's egg by the lower portion of the oviduct, or of the decidua of the mammalian fœtus by the uterus. It is more difficult to decide on the proper relations of the receptacles in those cases in which they exhibit spontaneous motions, or other indications of independent vitality. They may perhaps be considered in the light of special organs of the parent's body, destined to receive the corpuscules, and to be afterwards thrown off for their protection and dispersion, like the Hectocotylus-arm of the Cuttlefish, just referred to.

But such egg-sacs, or spermatophores, though in a certain sense distinct forms in the genetic cycle, intermediate between

---

* Siebold's Compar. Anatomy, I. § 290-348. Annals of Nat. Hist., 2d Ser., XVI., 150, XIX., 165. In some Crustacea the spermatophore seems to be formed by the coagulation of the outer film of the seminal fluid, on coming in contact with sea-water, the shape depending on that of a groove in the shell in which the spermatophore is modelled, as it were. Coste—Annals 3d Ser., II., 197. In *Geophilus*—a genus of Myriapoda — the impregnation seems to be effected by the female coming in contact with the spermatophore, which is suspended in the centre of a net stretched across one of the subterranean galleries of the insect. Annals, XIX., 165, 2d Ser.

† *Blatta* among Insects, *Purpura* and other genera of the Pectinibranchiate Mollusca, Nemertini, Hirudinei, and Lumbricini, among Annelida. The egg-sac of the earthworm (its so-called egg) contains two embryos—that of the leech as many as eighteen.

the parent and the embryo, cannot be considered as properly constituting a link of the series. However elaborate their organization, they are not really zooids, either protomorphic or gamomorphic, for they originate neither the embryos, nor the sexual elements of the embryonic existence; they serve merely to give a temporary protection to germs which are really of a date of formation anterior to their own. As the gamomorphic zooids have been compared to free spermaries and ovaries, so these receptacles may, according to the nature of their contents, be compared to spermathecæ, such as occur in insects, only free, or to free uteri, with broods of embryos enclosed—their office being not generation but gestation.

They do, however, simulate some of the more rudimentary zooids, both germinal and sexual, so that, under certain circumstances, they may be confounded, if the development be overlooked. Thus in the case of *Planaria* it appears still doubtful whether the bodies from which the broods of young issue are merely nidimentary envelopes of true ova, or germinal bodies, themselves *derived* from ova, and originating a fresh progeny by a proper act of gemmation.*

Many zooids, indeed, of both orders are known to naturalists, which, in a structural point of view, are little more than mere egg-sacs, and which in fact are, in some works, designated by this or an equivalent term, such as sporocysts, sporophores, sporosacs, &c. Such expressions, however, are, in strictness, open to objection, being in-

---

* Siebold's Comparative Anatomy, Ch. IX., § 129, with Dr. W. Burnett's note.

As Dr. Carpenter remarks, in discussing the embryogeny of *Purpura*, the development of *Planaria* may prove to be parallel to that of the Pectinibranchiate Mollusca, [noticed in Chap. III.] Trans. Micros. Soc., III., 17. See also Siebold, § 293, note 20, on the simulation of the parent cells of spermatozoa, by the spermatophores of some Crustacea.

consistent with precision of language, as well as with the true theory of development. A medusoid, for instance, is something more than an egg-sac or egg-carrier, it is an egg-producer, and the precursory form of a Trematode is neither one nor other—it is indeed a producer of its brood, not a mere carrier, but the brood consists neither of eggs, nor of embryos produced from eggs, but of zooids formed by a fresh act of gemmation.

Such objections are perhaps hypercritical, but the importance can hardly be over-rated, in a perplexed subject like the present, of avoiding the use of expressions which may in any way become a fresh source of confusion.

XI.

HOMOLOGICAL RELATIONS OF THE PARTS CONCERNED IN REPRODUCTION.

§ 1. IN order to complete our survey of the process of reproduction, it remains to offer some remarks on the correspondences or homological relations of the structures concerned in it, with reference to differences both of sex and species, and by way of making subsequent enquiries more intelligible, a few words may be premised concerning the principal modifications and general nature of the ultimate reproductive particles.

Having already adverted to the development of these corpuscules in both Kingdoms of Nature, the following summary of their principal modifications will be sufficient here, in connection with the remarks which have to be made on their essential nature, and on their homological relations.

The Spermatic Corpuscules in the Vegetable Kingdom are—

In the Protophyta, merely the condensed endochrome of certain cells, and undistinguishable from the germinal element.

In certain Algæ (Florideæ), Lichens and Fungi, ellipsoidal or cylindrical motionless bodies, without cilia *(microgonidia, spermatia, &c.)*

In the Algæ generally, ciliated ovoid bodies like minute zoospores *(antherozoids, phytozoa).*

In the Characeæ, and in the higher Cryptogamia generally, (Hepaticæ, Mosses, Ferns, Equisetaceæ, Lycopodiaceæ, and Marsileaceæ), filamentous motile corpuscules, somewhat resembling the spermatozoa of animals *(Antherozoids, Phytozoa)*.

In Phanerogamic plants, secondary cells formed in the anthers, and protruding, through openings in their outer or cellulose coat, tubules which are outgrowths from the inner coat, or rather from the protoplasmic granular matter which the latter encloses *(Fovilla* in *Pollen grains)*.

In the Animal Kingdom, as a general rule, these corpuscules appear to be developed as the nuclei of secondary cells *(vesicles of evolution)*, generated in variable numbers within larger parent cells. But in some Chilopoda, Acarina, and Entozoa, the wall, as well as the nucleus of the formative cells, seem to enter into their constitution; and in some Nematoid worms, and in Hydrozoa, it is alleged that they are formed directly as primary cells within the spermatic gland.*

The spermatozoa are generally motile filaments, often clavate—the larger extremity in some cases assuming the character of a distinct "body," rounded or acuminated—but in some Crustacea, in Chilopoda, and in Nematoid

---

* Huxley's Oceanic Hydrozoa, p. 22. The secondary cells are described by some as being formed by the breaking up of the large nucleus of the primary cell in a way somewhat similar to the cleavage of the vitellus of the ovum. It cannot be considered absolutely determined that the secondary spherules, within which the spermatozoa originate, are always true cells. See Burnett's note to Siebold's Comp. Anat. Invert., § 127. Quatrefages Ann. des Sci. Nat., 4th Ser., II., 152. The development of the spermatic particles in the higher Crustacea is described by H. Goodsir (Anatomical and Pathological Observations, p. 39), and more fully by Wagner and Leuckhart, as quoted below. According to the latter authors the spermatozoa are formed singly in projections from the primary cell [which may possibly represent the vesicles of evolution]. In the primary cell there is also an obvious nucleus, projecting externally, but it is eventually thrown off, without itself undergoing any farther development.

worms, they are motionless and of various forms, as speculate, clavate, elliptical, semi-lunar, &c. In the majority of animals these corpuscules are suspended in a peculiar seminal liquor, but this is wanting in many Insects, Worms, and other Invertebrata.*

The Germinal Corpuscules in Vegetables are—
> In the lower Cryptogamia the condensed granular contents of certain cells, which in the Protophyta are indistinguishable from the spermatic element.
> In the higher Cryptogamia the central corpuscules of the structures termed archegonia.
> In the Flowering plants, protoplasmic corpuscules lying within the embryo-sac, or central cavity of the ovule.

In the Animal Kingdom the almost universal arrangement is a nucleated cell, termed the *germinal vesicle*, occupying the interior of a larger one known as the *ovum*.

In both sexes there is reason to believe that the ultimate reproductive corpuscules are essentially mere homogeneous particles of the fundamental organic matter or protoplasm of the species, but so related to it and to each other, that neither is complete in itself, the conjunction of the two being required to constitute a true representative of the species. They are often termed cells, but a cell-wall is certainly not always present, and when present does not appear to be essential, but to serve only for the elaboration or protection of the real reproductive matter which it contains; for the true end of the latter cannot be attained till the impediment which the wall offers to the fusion of the corpuscules of opposite sexes is removed by its deliquescence, rupture, or penetration. This, indeed, is generally ad-

---

* Wagner and Leuckart, in Cyclopædia of Anatomy and Physiology, Vol. III., article, "*Semen.*"

mitted in the case of the spermatic particles, the only instance in which they are walled cells being that of the pollen grains of plants, the real value of which cannot be positively determined till we have greater certitude as to the behaviour of the extremity of the pollen-tube within the ovule. But the female germ is generally regarded as a true walled cell, and is considered moreover to be identical with the primordial cell of the future embryo, passing into it, when fertilized by some material or influence derived from the spermatic particles.

Yet the grounds are very slender for either of these assumptions—that is, either for admitting that the female corpuscule is—any more than the male—essentially a true walled cell, or that one of these corpuscules is more continuous than the other with the resulting embryo. The real state of the case appears to be that the embryo is formed out of the combination of masses of protoplasm representing the two sexes, and that it assumes afterwards the form of a true cell by its own progress of growth, the first steps of organization being—at least in animals—what is termed segmentation, and the formation of a wall round the exterior of the mass.

While it is true that these acts commonly take place within the cell-wall which originally surrounded the unimpregnated germ, this is not to be considered as an essential of their performance, but as due to the accidental circumstance that the commixture of the particles generally takes place by the spermatic corpuscules gaining access to the interior of this protecting cell, through a perforation in its wall; for when the commixture is effected otherwise the result is different. Thus, while in some of the conjugating Algæ the protoplasmic contents of one cell are transferred to another, with the contents of which they combine and form a spore to be matured in that same cell, in other species of the family *both* cells effuse their protoplasmic contents, and the spore resulting from their combination is either matured in a new provisional cell formed by an outgrowth from the walls of the conjugating cells, or is developed without the protection of any extrinsic cell-wall, though after a time it forms one for itself, by the induration of its exterior layer.

That in many at least of the higher Algæ, in which the two combining elements are differentiated into the representatives of sexual corpuscules, the germ is at first a mere protoplasmic mass, would certainly appear from the observations of Pringsheim, which go to show that it is not till after its combination with the phytozoa in the act of impregnation that a cell-wall is formed round it.* There is reason also to believe that the germinal bodies lying within the central cell of the archegonium of the higher Cryptogamia,† and the embryo-sac of the Phanerogamia,‡ are at first mere protoplasmic masses, which only after impregnation form on their exterior a true or distinct cell-wall.

In animals although in the reproductive organs of the female we have generally two well-marked cells, (the ovum and the germinal vesicle), one within the other, and each with a proper wall, neither seems to be the true homologue of the protoplasmic body of the vegetable, but rather the *contents* of the innermost—that is, the *macula* or nucleus of the germinal vesicle. Impregnation is effected by the spermatic particles diffusing themselves among the contents of the outer cell or ovum, either through a special aperture (micropyle) or in some of the other ways noticed before (Chap. III., p. 70) at the same time that the germinal vesicle dissolves and discharges its contents from within into the same cavity, so that the two sexual elements, lying

---

* Remarks on the spores of Algæ in Rep. of Berlin Academy (Quarterly Journal of Microsc. Science, IV., 126). Also Braun, Rejuvenescence in Nature (Ray Soc. Publ.), p. 156, and seq.

† So at least they would seem to be in *Buxbaumia*. See also Pringsheim, as quoted above.

‡ Griffith and Henfrey—Miscrograph. Dictionary, p. 432.

free within the ovum, readily coalesce. Then follows the cleavage of the yolk, and the formation of a wall on its exterior—the blastodermic vesicle.

§ 2. If we turn now to the homological relations of the germinal element in the two kingdoms of organic nature, and take for our starting point the protoplasmic corpuscules of the embryo-sac of the plant, and the contents of the germinal vesicle of the animal ovum, we may consider the "formative yolk" of the latter as the equivalent of the vegetable endosperm, and its primordial wall *(zona pellucida)* as representing the embryo-sac of the ovule of the Phanerogamia or the archegonium of the Cryptogamia. Among the adventitious parts or those met with only occasionally, we might compare the "nutritive yolk" of the bird's ovum, which is really derived from the contents of the ovisac, to the nucleus of the ovule or to the perisperm of the seed; and the albumen and shell of the egg (equivalent perhaps to the tunica granulosa and shaggy chorion of Mammalia) to the seed coats, arillus, &c. The ovary of the animal would represent the ovary or germen of the plant with its loculi.

In animals, on account of the germ not being generally—like that of the plant—impregnated *in situ*, that is to say, not coming in contact with the spermatic particles till it has escaped from the ovary, we have no aperture corresponding to the micropyle of the seed. The term "micropyle" has indeed come into use in connection with the animal ovum, but the aperture so called does not correspond to the micropyle of the seed, and is nowhere represented in the majority of plants, for though the canal of the archegonium of the Cryptogamia may fairly be taken to represent the micropyle of the egg in Fishes, Insects, Echinodermata, &c., there exists no such opening in the embryo-sac of the phanerogamic plant, the manner of fusion of the reproductive elements being as obscure as in the imperforate ova of the higher Vertebrata.

The oviduct—to complete the parallel—may be compared to the canal of the style in the flowering plant; while a certain analogy may even be suggested between the cleavage of the vitellus—giving rise to the cellular germ mass, which serves as the point of origin of the embryo of the animal—and the process of cell-multiplication by which is formed the suspensor, at one of whose extremities the embryo of the future plant is developed.

The following is a tabular statement of the correspondences now indicated; to push them farther would be a matter of pure fancy :—

Egg Structures.	Seed Structures.
Germinal Vesicle, with Macula.	Corpuscules of Embryo-sac, [Nucleus of Archegonium.]
Formative Yolk. [Cicatricula of Bird.]	Endosperm.
Primitive boundary of true ovum.	Embryo-sac, [Wall of Archegonium.]
Adventitious Yolk.	Perisperm.
Micropyle.	[Canal of Archegonium.]
Ovary and follicles.	Ovule, [Fern Spore.]
Oviduct.	Canal of Style.
The Adventitious Yolk and Vitelline Membrane of the Bird's egg are derived from the contents of the Vesicle.	The Perisperm and Seed-coats are derived from tissue and coats of the Ovule.

In regard to the homologues of the spermatic elements, the following is suggested as on the whole the most satisfactory parallel. Regarding the fovilla or contents of the pollen grain as the essential spermatic element, and therefore as the equivalent of the animal spermatozoon, the inner membrane of the pollen grain (intine) and its tubular outgrowth may be considered as adventitious structures—like the wall of the germinal vesicle—provided for the purpose of securing the advance of the effused fovilla in a regular column to its appointed destination, the embryo-sac of the ovule,* and may be regarded as homologous with the vesicle of evolution, whose nucleus in most animals becomes the spermatozoon. On this supposition the pollen grains would correspond to the parent cells within which the "vesicles of evolution" are generated, while the primary cells of the anther would represent the tubules, follicles, or other secreting cavities of the spermatic gland—relations which may be expressed in a tabular way, as follows:—

Vegetable Structures.	*Animal Structures.*
Pollen-tube and Fovilla.	Vesicles of evolution and Spermatozoa.
Pollen grain.	Spermatic parent cells.
Primary cells of Anther.	Seminiferous tubules, follicles, &c.
Anther.	Spermatic gland.

On the same principles the corresponding relations of the opposite sexual elements may be represented as follows:—

Animal Spermatic.	*Animal Germinal.*
Vesicles of evolution and Spermatozoon.	Germinal vesicle and Macula.
Parent Sperm-cell.	Ovum.
Spermary and follicles.	Ovary and Ovisacs.

* Darwin notices another provision for a like object in the long proboscidiform intromittent organ of some of the "parasitic males" among the Cirripedia. Speaking of this organ in *Cryptophialus minutus*, he remarks that its use "obviously is that the spermatozoa of these males, which are so extremely small in size, compared to the female, should all be conveyed within the sac, and none be lost. It should be borne in mind that the whole male, including every part, is scarcely larger than a single ovum, of which sometimes sixty have to be impregnated by only two or three males." The spermatozoa, he goes on to observe, have to pass a distance of about 1-7th of an inch from the gland to the lower ova—but of this about two-thirds would be along the canal of the intromittent organ (Monograph, II., 586). The envelopes of the female germ, in the Phanerogamia and Cryptogamia respectively, have already furnished us with a parallel case of the application of organs of very different homological import, to the discharge of similar functions.

Vegetable Spermatic.	*Vegetable Germinal.*
Fovilla within the Pollen-tube.	Protoplasmic Corpuscules.
Pollen grain.	Embryo-sac.
Primary Pollen-cells.	Nucleus of Ovule.
Parenchyma of Anther.	Ovule.

§ 3. By way of applying the principles of homological correspondence to the specialities of the several groups of organized beings, a few tabular statements are here subjoined, based on the notices of the reproductive process in Chapters II. and III. :

PROTOMORPHIC STAGE.

The primary product of fecundation, *(pro-scolex* passing into *scolex,)* is represented
    In Diatomaceæ, by the large frustules.
    In Algæ, Fungi, and Lichens, by the composite spores.
    In Mosses and Hepaticæ, by the seta and theca developed from the archegonial cell.
    In the higher plants, only by the cellular suspensor of the embryo of the seed.
It is represented in Animals by the primary germmass, which in many Invertebrata has the character of a ciliated animalcule. In general it attains no higher organization; but
    In Echinodermata it is developed into such zooids as *Auricularia, Pluteus,* and *Bipinnaria.*
    In Trematoda, into the primary gregariniform zooid *(redia* or *sporocyst).*
    In Cestoidea, into the *Cysticercus* or the primary vesicle of the *Echinococcus.*

The second and following intermediate forms *(meta-scolex)* are represented

> In Mosses, by the spores, and filaments of the protonema into which they germinate.
> In Trematoda, by the second and subsequent gregariniform zooids.
> In Cestoidea, by the secondary and subsequent vesicles of the *Echinococcus*.
> But in the majority of Plants and Animals, not at all.

ORTHOMORPHIC STAGE.

The rudimentary condition of the typical form is represented in plants and animals generally by the embryo.

> In Fungi, by the confervoid filaments of the nascent mycelium.
> In Lichens, by those of the hypothallus.
> In Echinodermata, by the "disc."
> In *Ascidium*, by the *Spinula*.
> In *Distoma*, by the *Cercaria*.
> In Insects, by the "maggot" or "caterpillar."
> In Batrachia, by the "tadpole."

When naked, the embryo is termed a *larva*; when extruded from the body of the parent, but still invested with its own envelopes, it is termed a *seed* or *egg*.

Secondary and subsequent phytoids or zooids, more or less of the typical form, are represented—

> In Vegetables, by the successive pullulations of adherent leaf-shoots, making up the ramose vegetation of the majority of plants.
> In the Polypifera and Polyzoa, by the successive polypes, which, with their connecting stalks, form the polypidom.
> In Annelida, by the "joints" of the body.

In Insects, by the "viviparous *Aphides*" and other like forms.

The ultimate phase of the typical stage is represented-
In Fungi, by the fruit-bearing plant.
In other lower Cryptogamia, by the thallus.
In Mosses, by the perichætial leaves.
In Ferns, Equisetaceæ, Lycopodiaceæ, &c., by the sporangium.
In the higher plants, by the floral organs.
In the Polypifera, by the "urn" *(gonophore)* and blastostyle.
In the Trematoda, by the Distoma-form.
In the Cestoidea, by the mature Tænia-head.
In Insects, by the Imago, generally winged.
In Animals generally, by what is recognised as the *adult form.**

GAMOMORPHIC STAGE.

The proper sperm or germ stock *(strobila, proglottis)* is represented—
In Algæ, by the androspore of *Bulbochæte*.
In Fungi, by the hymenium.
In Lichens, by the apothecium.
In Mosses, scarcely at all.
In the higher Cryptogamia by the spore, and the prothallium developed from it.
In the Phanerogamia, by the pollen grain and ovule.
In the Polypifera, by the medusoid, or the sporosac.
In the Cestoidea, by a proglottis, or cucurbitiform segment, as in *Tænia*.

* In the higher Animals, however, as will be presently noticed, the adult state represents, in addition to the typical form, the rudiments also of the gamomorphic, in the parts of sex, implanted at a comparatively early period of embryonic development, on the pre-existing structure known as the *Wolffian body.*

In the Tunicata, by a catenated zooid, as in *Salpa*.

In Annelida, by a caudal zooid, as in *Syllis*.

In Animals generally, by the spermary and ovary.

The sperm-sacs and germ-sacs are represented—

In the Protophyta, by the conjugating cells.

In Algæ, by the antheridium and the sporangium.

In Fungi and Lichens, by the spermagonium and ascus.

In the higher Cryptogamia, by the antheridium and archegonium.

In the Phanerogamia, by the pollen-tube, and the embryosac of the ovule.

In Animals generally, by the primary sperm-cell and ovum.

The ultimate sexual elements are represented—

In the Protophyta, by the endochrome of the conjugating cells.

In Algæ, by the antherozoids and germinal corpuscules.

In Fungi and Lichens, by spermatia and endothecal nuclei.

In the higher Cryptogamia, by the filaments in the cellules of the antheridia, and the central corpuscule of the archegonium.

In the Phanerogamia, by the fovilla in the pollen-tube, and the germinal corpuscules of the embryosac.

In Animals generally, by the spermatozoa in the vesicles of evolution, and the macula of the germinal vesicle.

Tabular views are given in the Appendix of the points of correspondence as above indicated.

§ 4. In conclusion, a few observations may be made on the relations of the sex to the individual.

If this term *individual* be applied to every isolated organism, then it is only in cases in which—as in the higher species—no process of detachment of gemmæ is interposed, at any period of the life-history, so as to interfere with the integration into a single body of the successive phases of organization, that the character of sexuality can attach to every individual of the species. On a first view, indeed, its limits appear even more restricted, for, though in the higher animals—except in circumstances of immaturity, or of mutilation, or other abnormality—all the individuals have true sexual powers, yet, in certain families of the Invertebrata—as Bees and Termites, among insects—the reproductive faculty is not diffused over the whole species, the majority being destitute of all power, either of impregnation, or conception, and hence termed *neuters*, as being neither males nor females. This sterility, however, is rather functional than organic, the so-called neuters being, in fact, females, in which the reproductive organs have remained in a rudimentary condition.

Generally in the animal kingdom, the development is one-sided, the same indvidual not producing both sexual elements. In the normal condition of the higher animals, we never find the sexes united, each individual being either simply male or simply female. This applies universally to Vertebrata, except as a rare monstrosity.* It applies also very generally to the typical Articulata, except in aberrant orders, as the Cirripedia, but in the vermiform tribes, and in the sub-kingdom of Mollusca, the union of the sexes, by what is termed a hermaphrodite or bi-sexual arrangement, is common enough.

---

* It seems to occur more frequently in fishes than in any of the higher classes, and is said to be a normal phenomenon in the perch.—(Dufossé Annales des Sciences Naturelles, 1857.)

With these qualifications, it still remains true that where the unity of the organic structure is not broken in upon by an interpolated process of gemmation, all the individuals of the species, as their normal constitution, present indications of sexuality. They may not all be endowed with reproductive power, but they all normally possess organs, either male or female, or both combined.

It has been contended, indeed, that the original uniformity of type is even more absolute than this—that sexual organs, of both kinds, exist at first normally in all the individuals of a species, though the full development of one sex is usually associated with the non-development of the opposite—a view originally suggested by Dr. Knox,* and one which has several theoretical considerations in its favour. It goes far to harmonize the existence of the sexual peculiarities of individuals with the general laws of reproduction, in those species whose continuity of organization is not broken up by any separation of free zooids; for it certainly seems more natural in this case to regard all the individuals as having an original structural identity, and to ascribe the sexual differences to the non-development of certain parts, than to suppose the admitted rule of like producing like to be continually broken in upon by essential differences of organization from the very first in the parts of generation. It also serves to soften the abruptness of the contrast between the normal hermaphroditism of some species, and the arrangement of separate sexes, which is the ordinary rule in the animal kingdom. And it affords, farther, a ready explanation of the occasional appearance of hermaphrodites, even in species in which the sexes are normally separate.

This is, indeed, a rare occurrence, for the individuals so termed are commonly only malformed males or females. But a monstrosity, consisting of a real intermixture of sex,

---

* Brewster's Edinburgh Journal of Science, II., 322—1830.

does at times occur, and may, on this view, be explained in accordance with the general principles of teratology. Congenital malformations are generally the result either of defect or excess of development, or of the irregular association of both influences; and the latter is exactly the condition calculated to induce such a monstrosity, on the hypothesis in question. The most common form of intermixture is probably the development of male organs on one side, and female on the other—a monstrosity which has been observed in Fishes and Insects.* In the latter, it reveals itself by a corresponding disparity in the external characters of sex, as in the markings of the wings of Butterflies.† But there are not wanting instances of a true intermixture of male and female organs on the same side of the body.‡ Such cases afford a strong argument in favour of the original co-existence of the organs of both sexes as a normal arrangement, for otherwise we must first assume the co-existence of another malformation—viz., the duplication of the organs on one or both sides—and then their sexual development in opposite ways; a view which is quite opposed to the general rule, that in double monsters the two individuals are always of the same sex.

It does not appear, however, that the speculation of Dr. Knox is borne out by the embryonic development of the parts in the higher animals, in which alone the point has been carefully examined. It is true it cannot be said either to be *disproved* by this test; only there is an entire absence

---

* Siebold's Comparative Anatomy, § 318, note 1.
Dr. Simpson, on Hermaphroditism. Obstetric Works, Vol. II., pp. 242-304—a reprint, with some additional cases and remarks, from an article in the Cyclopædia of Anatomy and Physiology, Vol. III.

† An illustration of this bi-sexual livery is given in Humphrey's Butterfly Vivarium, page 142, and plate II. fig 6.

‡ Simpson Op. Cit., pp. 280-339.
Dr. Bluckman and Dr. W. J. Burnett, in the American Journal of the Medical Sciences, N. Ser., Vol. XXVI., pp. 63-367.

of any evidence in support of the co-existence of the *essential* organs of the two sexes—the spermatic and ovarian glands—which is the more remarkable from there being a very complete representation of the *associated* structures, both male and female.*

The organs of both sexes are developed on the basis of prior embryonic structures, the Wolffian bodies. In connection with these the embryo exhibits on each side two hollow filaments or tubules—the efferent duct of the Wolffian body and the Mullerian cord. According to Kobelt, in the male the former is converted into the vas deferens, while the latter almost disappears. In the female, the latter is converted into the fallopian tube, while, in the human species, the duct entirely disappears, though in Ruminantia and Pachydermata, it remains as the rudimentary structure known by the name of the canal of Gartner.† Morphologically, the uterus is merely a dilatation of the common inferior termination of the two oviducts or fallopian tubes, and is represented in the male by the prostatic vesicle, which, though a mere follicle in the human species, assumes, in some of the lower animals, so much the character of a miniature uterus, with two filaments or tubules going off from its upper angles, alongside the vasa deferentia, that Weber gives it the name of "Uterus Masculinus."‡ Up to a certain period of

---

* This correspondence becomes still more striking, if we extend our survey to the other Mammalia, in some of which we find the scrotum cleft, in others the clitoris traversed by the urethra. Simpson on Hermaphroditism, Obstretic Works, Vol. II., pp. 220-307. It is to an abnormal development of these accessory parts that the multiform disguises of sex are due, which are popularly confounded with true hermaphroditism.

† Valentin's Text-book of Physiology, 661.

‡ Zusätze zur Lehre vom Bau and den Verrichtungen der Geschlechtsorgane. Leipsic, 1846.

The development is well marked in the Beaver, Rabbit, Boar, Horse, Ass, Badger, Goat, Sheep, and Deer.

But, perhaps, as is pointed out by Leuckart, the part may be more ac-

embryonic development, both sets of organs continue equally apparent, so that the sex is with difficulty distinguishable, as is particularly remarked by Weber in the case of the Rabbit.

The Wolffian body itself, there is reason to believe, has a much more permanent existence than was at one time supposed, for a structure, which is probably to be regarded as its residuum (parovarium) has been discovered by Rosenmüller in the broad ligament of the uterus of the fœtus and infant,* and has been shown by Kobelt to continue to exist in the same situation, even in the adult female,† while, in the male it is converted into the epididymis, and, perhaps, in part (superiorly) into a peculiar body, to which Giraldès has called attention, in the spermatic cord *(Corps Innominé.)*

So far, therefore, as appears from the facts of development, we may either assume, in conformity with Knox's view of the primordial duplicity of sex, that in the whitish spot on the inner margin of the Wolffian body, where the

---

curately compared to the whole genital sinus of the female—that is, to the uterus and vagina taken together. (Cyclopædia of Anatomy and Physiology, Art. *Vesicula Prostatica*, Suppl., p. 112.) The Prostrate Gland, in which it is imbedded, has been held to represent the glandular substance of the cervix uteri, but this view is negatived by its occasional occurrence conjointly with a distinct uterus. (Dr. Warren, in Amer. Journ. of Med. Science, July, 1857, p. 127.) Its homologues, therefore, as well as those of the vesiculæ seminales, are somewhat uncertain. Dr. Simpson suggests that the former may be represented by the follicles of the walls of the female urethra, while the latter may be regarded as analogous appendages of the seminal canals. The relations of the external organs are more obvious, but it is not necessary for our subject to enter upon them. A full tabular view of the homologies of all the parts, both external and internal, is given in Dr. Simpson's article on Hermaphroditism, as reprinted in the second volume of his Obstetric Works, p. 320.

* "Quædam de ovariis embryonum et fœtuum humanorum."

† Philosophical Transactions, April, 1858. See also Dr. A. Farre, in Supplement to the Cyclopædia of Anatomy and Physiology, Art. "Uterus," p. 593.

generative gland first appears, there are distinct foci of development for both sexual elements—though the normal development of one suppresses that of the other;—or we may adopt the more common theory, that there is but a single focus of this kind, whose development is, in some unknown way, directed exclusively to the evolution of one or other sex. The latter view, certainly, is that most naturally suggested by the facts of the case in the higher species, and also by the development of the gonophores of the Hydrozoa. In the lower animals, generally, the primordial condition of these organs has not yet been much studied in this point of view.

When the sexes are separate, two individuals—male and female—are required for an adequate representation of the species; but, in cases where they are united, each individual of the species becomes a complete type of the whole, so far as structure is concerned. Functionally, however, it is only so, when the bi-sexual organism has the power of self-impregnation; and this is comparatively seldom the case, most hermaphrodites impregnating each other, either by a reciprocal act (as in *Helix*), or by their aggregation into chains, the individuals composing which receive fecundation from those in advance, and communicate it to those behind.*

On the other hand, in the cases in which the oneness of the organism is broken in upon by an interpolated process of gemmation, the sexual characters attach themselves ex-

* In some bi-sexual Cirripedes, it has been shown by Mr. Darwin that a sort of supplementary impregnation is effected by separate males of microscopic size and parasitic habits, which he terms complementary males.

In aquatic animals, intromission of the spermatic particles is by no means so regular a phenomenon as in terrestrial species, impregnation being frequently effected by the diffusion of the seminal fluid through the water in the neighbourhood of the previously deposited spawn of the female. The process may be imitated artificially, and this is now done to some extent in the breeding of fish.

clusively to those of the resulting zooids or phytoids, which represent, or are more or less incorporated with that phase of development to which, in the foregoing pages, the term gamomorphic has been applied.* The other links, which help to make up the entire series intervening between the successive acts of sexual reproduction, are truly neuter— that is, not only incapable of exercising sexual *functions*, but destitute of all trace of sexual *organs*, and multiplying simply by some modification of the budding process.

Protomorphic organisms thus isolated from those of the succeeding phases have not yet been shown to have their neutrality even indirectly affected by their giving rise, in individual cases, exclusively, to male or female gamomorphic forms ; but analogous differences have been satisfactorily determined as occurring in the orthomorphic or typical phase. In the vegetable kingdom particularly, they have long attracted the attention of botanists, who apply the term *diœcious* to species whose sexual phytoids or flowers are exclusively of one sex on any single plant, and that of *monœcious* to species whose sexual phytoids, though themselves exclusively either male or female, occur of both kinds in connection with the same plant. A word of the same general character—such as *synœcious*—might be devised to distinguish plants bearing bi-sexual flowers, to which as yet no special term has been applied. The expressions might farther be extended with advantage from phanerogamic plants, not only to those organisms—like Mosses, and some of the Polypifera, in which the reproductive organs are in obvious connection with the vegetative stock,—but to other cases also, in both kingdoms of nature, in which the sexual forms are quite isolated. In the case of Ferns or Polypifera, for instance, we should call a species synœcious, in which the prothallia or medusoids are bi-sexual; another

---

* Nurses *(Amme)* of Steenstrup ; Agamozooids of Huxley.

monœcious, in which they are of separate sexes; and another, again, diœcious, in which those of one sex only are derived from the same axis or polype stock. It is believed that illustrations of all these arrangements may be found in the two groups referred to, as well, perhaps, as in some others.

The sexual relations, now indicated, may be represented in a tabular form as follows:—

Reproductive Organs present, as integral parts of the system :—
    Male and Female in different individuals, each of which is............... Monosexual.
    Male and Female in the same individual, which is, therefore............ { Bi-sexual or Hermaphrodite,
    but not necessarily self-impregnating.
Organs appearing as distinct gemmæ :—
    Male and Female in the same gemma... Synœcious.
    Male and Female in different gemmæ, from the same stock.................. Monœcious.
    Male and Female in different gemmæ, from different stocks.................. Diœcious.
Organisms, normally incapacitated for sexual action, are termed........................ Neuter.
    True neuters occur only in the preliminary phases of species which have the sexual organs in detached gemmæ;
    False neuters are either rudimentary or larval conditions of organisms which afterwards form true sexual organs;
    Or, individuals in which the latter remain permanently in a rudimentary condition, as in the female working Bees.

§ 5. It is with some hesitation that I have used the term homologous with reference to the correspondences above suggested, as they are obviously of a different nature from those to which the expression was originally applied by Professor Owen. In comparative anatomy, the term homologous is employed to indicate parts which occupy corresponding positions in that general plan of conformation traceable throughout large groups of animals. Such parts have no farther community of office than arises from their physical adaptation. The work which they are naturally adapted to perform, in the more typical species, they must be suited for in all, in some degree, so long as there is no wide departure, on the one hand, from the ordinary form of the organ, or, on the other, from the ordinary habitat of the group of animals. But, when this does occur, we have numerous examples of their being turned to totally different uses, or left in a rudimentary state, without any office to perform at all. It is sufficient to cite, in illustration, the rudimentary condition of the pelvis of the whale, or the transference of the respiratory function in the fish from the lung to the gill, the rudimentary homologue of the former organ being left to do duty as a swimming bladder.

But, in the correspondences which have been under consideration in this chapter, the function or office to which the parts minister is a point of prime importance in estimating their relations. The spermatic and germinal organs, for instance, have throughout been assumed as the corresponding parts in the economy of the two sexes. Yet, there is reason to believe, they are not always really structural homologues ;* and all that is meant by instituting a comparison between them, is to indicate that they occupy corresponding or equivalent positions in the reproductive process, in as much as they are similarly related to the orthomorphic phase of development on the one hand, and to the ultimate spermatic and germinal corpuscules on the other. Hence, it may appear that the term *analogous*

* See Sanderson in Supplement to Cyclop. Anat. and Physical " Veget. Reproduct.," p. 252.

would be more suitable, as now commonly restricted to correspondences in function. There is a wide difference, however, between the reproductive function and others, in the constancy of its structural relations. Whether it is that the process, whereby are laid the foundations of the organized fabric, has, as it were, reflected back upon it some of the uniformity of system, which is so striking a characteristic of the latter, or whatever other explanation may be suggested, so it is, as a general rule, that not only have the reproductive organs as definite relations in each great group as any part of the fundamental framework of the skeleton, but that even in animals of totally different types of organization, there prevails a close uniformity of structure, even in very minute details of these parts, the evolution, both of ova and spermatozoa, presenting—with some few exceptions—the closest resemblance throughout the whole Animal Kingdom.

It is true that the interpolated process of gemmation observes no such constancy, so that the uniformity is very much confined to the gamomorphic or sexual stage; but as the phenomena of this stage form the starting point of the cycle of reproduction, the use of the term *homologous* appears to indicate, at least, the general character of the process, better than any other in common use.

I have only, in conclusion, to repeat my regret, if the spirit of system has in any respect given a false bias to the interpretation of the facts of nature. Impressed, as I have long been, with the conviction of the prevalence of an orderly arrangement throughout the works of the great Creator, who "hath made all things double one against another;" and, believing that some such guiding principle has been successfully used by the most cautious explorers of nature, as a clue to her more intricate mazes, I freely admit having proceeded, in the examination of the comparative details of the process of reproduction, with a strong prepossession in favour of the existence of a real, though hidden unity, in the essentials of this important function throughout the whole range of organised beings. Feeling, however, at the same time, that the very consciousness of such a prepossession is some protection against its leading into serious error, I now submit my conclusions, such as they are, for the judgment of those whose position among the cultivators of Natural Science entitles them to pronounce upon them.

APPENDIX.

It appears the most convenient arrangement to put the following Tables in an Appendix by themselves, as referred to in different parts of the work.

Tables I.—IV. are slightly altered from those appended to my paper on the Genetic Cycle, in the Edinburgh Philosophical Journal for January, 1860. Table V. is a modification of one given by Dr. Sanderson, in the Cyclopædia of Anatomy and Physiology (Art. Vegetable Ovum); the explanation appended to it shows in what respects his views have been departed from.

The three succeeding give a conspectus—arranged as much as possible in the same plan, for the sake of comparison—of the theory of relations which I have been led to adopt in this work.

GENETIC CYCLE IN ANIMALS.

ORDINARY REPRODUCTION.

PROTOMORPHIC STAGE.

- Development of the Cellular Germ-Mass or "Mulberry Body."
- Budding off of the "Primitive Trace" of Embryonic Organization.

ORTHOMORPHIC STAGE.

Development of the Typical Form,

- through the successive phases of Embryogeny; (Metamorphosis in the case of Larvæ or Naked Embryos, as in some *Crustacea*, *Insecta*, and *Batrachia*.)
- Formation of the matrix of the Ovary and Spermary (Wolffian Body.)

GAMOMORPHIC STAGE.

- Development of the proper Ovarian and Spermatic Follicles.

FECUNDATION OF THE OVUM.

ALTERNATION OF GENERATIONS.

- Development in *Echinodermata*—of *Auricularia*, *Bipinnaria*, *Pluteus*; in *Cestoidea*—of Cystic Forms; in *Trematoda*—of Infusorian, and Gregarinic Forms.
- Pullulation of Derivative Protomorphic Zooids, as in *Echinococcus*, some *Trematoda*, &c.
- Budding off of the Disc of the *Echinodermata*, or the Cercariform Embryos of the *Trematoda*.
- as, in *Echinodermata*, of the Starfish or Sea Urchin; in *Cestoidea*, of the "Tænia-head;" in *Trematoda*, of the Cercariform Larva and the *Distoma*.
- Pullulation in *Polypifera* and *Polyzoa*, of a series of Polypes cohering as a Polypidom; in *Aphis*, of successive swarms of free Larvæ, like the original.
- Sprouting of the "Gonophores" of the *Polypifera*; or the "Stolons" of the *Tunicata*.
- Development in *Polypifera*, of Medusiform Zooids; in *Salpa*, of the Catenated Form; in *Annelida*, of Caudal Zooids; in *Cestoidea*, of Proglottides.
- Pullulation of Secondary Medusoids in *Sarsia* and some other Species.

Formation of the Reproductive Elements, i.e., Ova and Cells with Spermatozoa.

PHANEROGAMIA.

GENETIC CYCLE IN PLANTS.

PROTOMORPHIC STAGE.

Primary Development from the Fecundated Germ,

in *Hepaticæ* and *Mosses*, of the Theca, and Spores, which of the Cellular Pro-embryo or Suspensor.
in the latter germinate into Protonemic Filaments
- Pullulation of Derivative Protonemic Filaments; — Quaternate division of the Suspensor in *Coniferæ*.
- Formation of the Gemma of the leafy axis of *Mosses*; — Formation of the Embryonic Cellular Mass.

ORTHOMORPHIC STAGE.

Development of the Typical Form or Vegetative Axis of the
Moss, Fern, or Flowering Plant.

Pullulation of Successive Shoots, generally remaining in adhesion, but sometimes developed from deciduous bulbs.

Formation of the Perichætia of *Mosses*, — Formation of the Floral Organs.
or the Sporangia of *Ferns*;

GAMOMORPHIC STAGE.

Development of the proper Reproductive Structures, viz.:—

of the Spores of *Ferns*; — of Ovules and Pollen-grains.

Pullulation of the Prothallium from the Fern-Spore, and of the "albuminous body" in the Ovule of *Coniferæ*.

Formation of the Reproductive Elements, viz.:—

{ of Archegonia, with Germ-cells { of Embryo-sacs with Germinal Bodies
  and Antheridia, with Cellules containing Antherozooids and Pollen-tubes with Fovilla
    (or in some *Coniferæ*, with Secondary Cellules.)

FECUNDATION.

CRYPTOGAMIA.

TABLE III.

PERIODS OF INTERPOLATION OF GEMMATION IN THE GENETIC CYCLE

IN SEVERAL TRIBES OF PLANTS AND ANIMALS.

PROTOMORPHIC.	ORTHOMORPHIC.	GAMOMORPHIC.
Mosses & Hepaticæ.	Phanerogamia	Ferns & Equiseta.
Echinodermata (Polyzoa) Trematoda Cestoidea (Annelida)	Polypifera Polyzoa (Annelida) Aphides	Polypifera Polyzoa Salpæ Cestoidea Annelida *(Syllis, &c.)*

Gemmation is exceptional in any stage among the higher Articulata and Mollusca, and is unknown, as a normal arrangement, among Vertebrata.

TABLE IV.

RESTING PERIODS IN THE GENETIC CYCLE.

Mosses and Hepaticæ *protospores* } In the middle of the Protomorphic stage.

Ferns and Equiseta *gamospores* } Between the Orthomorphic and Gamomorphic.

Phanerogamia *seeds* } Between the Protomorphic (embryogeny) and the Orthomorphic (vegetation commencing with germination).

Coniferæ present also another resting period, in the middle of the Gamomorphic stage (during the maturation of the fruit).

Animals in general, *eggs* or *ova*, } Between the Gamomorphic (maturation of the reproductive organs), and the Protomorphic (embryogeny).

Insects, Trematoda, &c., } Have also a resting period (*pupa* or *encysted state*) during their metamorphosis, in the early part of the Orthomorphic stage.

Mammalia have no obvious resting period, the mature ova requiring immediate fecundation, which is at once followed by segmentation and the development of the embryo.

The bodies which, under the names of *statoblasts, bulbs, resting spores*, &c., perform the part of eggs or seeds in some species of animals and plants, appear occasionally to be gemmæ, which may be termed *accessory*, as lying out of the direct genetic cycle.

TABLE V.

Abstract of Dr. Sanderson's Table of the Analogies

Algæ, Fungi, Lichenes.	Hepaticæ, Musci.	Filices, Equiseta.	Rhizocarpeæ.
SPOROID BODY Germination Conferva, Mycelium, or Hypothallus	SPORE-CELL G. into a Filamentous Protonema	SPORE-CELL G. into an Exosporous Prothallium	MACROSPORE G. into an Endosporous Prothallium MICROSPORE G. by the protrusion of the Inner Spore-coat
Thallus	Frond or Leafy Stem	Prothallium (Bi-sexual)	Prothallium (Female)
Nucule (of Chara) GERM-CELL Globule (of Chara) SPERMATIA, &c.	Archegonium GERM-CELL Antheridium ANTHEROZOIDS	Archegonium GERM-CELL Antheridium ANTHEROZOIDS	Archegonium GERM-CELL Absent ANTHEROZOIDS
	Division of Germ-cell From lower segment FRUIT-STEM	D. of Germ-cell Formation of EMBRYO	D. of Germ-cell Formation of EMBRYO
	SPORE-BEARING STEM Sporangium Spore	SPORE-BEARING PLANT Sporangium Spore	SPORE-BEARING PLANT Sporangium Spore and Microspore

TABLE V.

in the Development of the Different Classes of Plants.

Lycopodiaceæ.	Gymnospermous Phanerogamia.	Angiospermous Phanerogamia.
MACROSPORE G. into an Endosporous Prothallium	ALBUMINOUS BODY	EMBRYO-SAC
MICROSPORE *G. by the protrusion of the Inner Spore-coat*	POLLEN-CELL *G. by the protrusion of the Pollen-tube*	POLLEN-CELL *G. by the protrusion of the Pollen-tube*
Prothallium (Female)	Absent	Absent
Archegonium GERM-CELL *Absent* ANTHEROZOIDS	Corpusculum GERM-CELL *Absent* *Absent* [FOVILLA]	Absent GERM-CELL *Absent* *Absent* [FOVILLA]
D. of Germ-cell From upper segment Suspensor, From lower segment EMBRYO	D. of Germ-cell From upper segment Suspensor, From lower segment EMBRYO	D. of Germ-cell From upper segment Suspensor, From lower segment EMBRYO
SPORE-BEARING PLANT Sporangium Spore *also* Microsporangium Microspore	OVULIFEROUS PLANT Ovule Albuminous Body ANTHERIFEROUS PLANT Anther Pollen	OVULIFEROUS PLANT Ovule Embryo-sac ANTHERIFEROUS PLANT Anther Pollen

In the foregoing Table the embryo-sac of the higher Phanerogamia is taken as representing the "albuminous body" of the Coniferæ—or rather its wall, for its cellular contents would answer more to the endosperm. Hofmeister farther considers this body to represent the prothallium of the Cryptogamia.

The view here adopted is distinct from both, regarding the following as homologous plants :—

1. The spore and the ovule.
2. The prothallium and the "albuminous body."
3. The Archegonium, the Corpusculum, and the Embryo-sac.

In this view the endosperm is unrepresented in the Coniferæ and Cryptogamia, and the prothallium and "albuminous body" in the Angiospermeæ.

But the principal point, in which the relations represented in this table differ from those advocated in the present work, is the identification of the alternation of the Mosses with that of the Ferns—as has been explained before at some length.

To make the points now referred to more intelligible I have represented in a tabular form the relations here adopted, breaking up Dr. Sanderson's Table into three, for the sake of convenience, but keeping to his general arrangement, as much as the necessary changes allow. In these tables, as well as in the one just given, the male structures are printed in italics.

TABLE VI.

TABLE OF THE ANALOGIES IN THE GENETIC CYCLE OF THE LOWER CRYPTOGAMIA.

Structures.	Algæ, Fungi, Lichens, &c.		Hepaticæ, Musci.	
Protophyte	Compound Spore (Xenodochus)		Fruit-stem	
Protospore	Uredo-spore (Æcidium)		Spore	
2d Protophyte			Protonema	
Embryo	Mycelium		Nascent Moss	
Typical form	Thallus		Moss-stem	
	Pullulation of Fronds		Pullul of Shoots	
Matrix of Reproductive organs	Hymenium		Perichætium	
	(F)	(M)		
Gamospore	Secondary Spores	Microgonidia (Œdogonium)	absent	
Gamophyte	Gynothallium (Ceraminm)	Androthallium (Œdogonium)	absent	
			(F)	(M)
Sex Capsule	Ascus (Fungus, Lichen)	Spermagonium (Lichen)	Archegonium	Antheridium
	Nucule (Chara)	Globule (Chara)		
Sex Elements	True Spores	Spermatia	Germ-cells	Antherozoids
	FECUNDATION.			

TABLE VII.

TABLE OF THE ANALOGIES IN THE GENETIC

Structures.	Filices, Equisetaceæ.	
Protophyte	...Suspensor	
Protospore..............	...absent	
Embryo...................	...Embryo	
Typical Form	Fern-stem and Fronds................	
	Pullulation of Shoots..................	
Matrix of the Reproductive Organs	Sporangium	
Gamospore	Spore germinating into a	
	(F)	(M)
Gamophyte	Prothallium	*Prothallium*
Sex Capsule	Archegonium	*Antheridium*
Sex Elements	Germ-cell	*Cellules with Antherozoids*
	FECUNDATION.	

TABLE VII.

CYCLE OF THE HIGHER CRYPTOGAMIA.

Rhizocarpeæ.		Lycopodiaceæ.	
...Suspensor ...absent		...Suspensor ...absent	
...Embryo ...Leafy Axis ...Pullulation of Shoots		...Embryo ...Leafy Axis ...Pul. of Shoots and Spikes	
...Sporangium		(F) ...Oophoridium	(M) *Androsporangium*
(F) Macrospore	(M) *Microspore*	Macrospore	*Microspore*
Prothallium	*absent*	Prothallium	*absent*
Archegonium	*absent*	Archegonium	*absent*
Germ-cell	*Cellules with Antherozoids*	Germ-cell	*Cellules with Antherozoids*
FECUNDATION.			

TABLE VIII.

TABLE OF ANALOGIES IN THE GENETIC CYCLE OF THE PHANEROGAMIA.

Structures.	Gymnospermeæ.		Angiospermeæ.	
Protophyte..	...Suspensor		...Suspensor	
Protospore...	...Absent		...Absent	
Embryo	Embryo		Embryo	
Typical Form	Female and Pullulation	*Male Plants* of Leaf-shoots	Female and Pullulation	*Male Plants* of Leaf-shoots
Matrix of the Reproductive Organs	Ovuliferous Flower	*Antherif. Flower*	Ovuliferous Flower	*Antherif. Flower*
Gamospore	Ovule	*Anther*	Ovule	*Anther*
Gamophyte	Albuminous body	*Pollen-grain*	Absent	*Pollen-grain*
Sex Capsule	Corpusculum	*Pollen-tube*	Embryo-sac	*Pollen-tube*
Sex Elements	Germ-cell	*Cellules with Fovilla*	Germ-cell	*Fovilla*
	FECUNDATION.			

The three foregoing Tables (VI., VII., VIII.,) are in substance the same as Tab. I., but the differences in arrangement allow some points to be brought out more clearly—such as the extent to which the sexual characters affect the sequence of forms. Omitting the Thallogenous group—of which our knowledge is still too fragmentary and conjectural to allow us to generalize upon it—there will be seen to be a progressive increase in the extent of the genetic cycle, over which this influence prevails, as we rise from the lower Cryptogamia to the Phanerogamia, and especially to the dioecious species of the latter. The difference of manifestation is even greater than can be measured in this way, as the sexual characters of the lower forms are not only more limited in extent, but also of a less apparent kind—fully justifying the expressive term given by Linnæus to the lower of the two great divisions of the Vegetable Creation.

GENETIC CYCLE IN ANIMALS.

For the Animal Kingdom, on account of the great variety in the Genetic Cycle, it would be difficult to construct convenient and manageable tables arranged in parallel columns like the foregoing, but I have appended in their stead tabular views of the sequence of development in the families which are of most interest in respect of their mode of propagation.

In the following Tables (IX.-XVIII.), the division into columns indicates the existence of sexual distinctions in that part of the cycle where it occurs—the left-hand column giving the spermatic, and the right, the germinal structures. Unless contra-indicated by the punctuation, the reading is to be carried across the page. The horizontal lines indicate a breach in the continuity of structure. This may sometimes affect only one sex, as in the detachment of the spermatozoa of animals generally, while the ovum is commonly impregnated, and often incubated, within the body of the female.

Brackets are used to imply that the parts or processes mentioned are only of occasional occurrence in the group.

TABLE IX.

GENETIC CYCLE OF THE POLYPIFERA.

PROTOMORPHIC STAGE,
(merged in the following).

The ciliated Germ—*(planula)*.

ORTHOMORPHIC STAGE.

The Polype.
[The Ramifications of the Zoophyte].
The gemmiparous Vesicles [*gonophores* of Allman].

GAMOMORPHIC STAGE.

Medusoids or Sporosacs,
[and Secondary Medusoids and Sporosacs budded off from these,]

(male and	female.)
Spermatic, and	Ovigerous Sacs.
Primary Sperm Cell ?	The Ovum (unwalled).
Special Sperm Cells ? and Spermatozoa.	{ The Germinal vesicle and Macula.

FECUNDATION.

TABLE X.
GENETIC CYCLE OF THE ECHINODERMATA.

PROTOMORPHIC STAGE.

The ciliated Germ, passing into the condition of an *Auricularia*, *Bipinnaria*, or *Pluteus*.

ORTHOMORPHIC STAGE.
The Echinoderm,
(male and female).

GAMOMORPHIC STAGE,
(merged in the foregoing).

Spermary,	Ovary.
Primary Sperm Cell,	Ovum.
Special Sperm Cells and Spermatozoa.	The Germinal Vesicle and Macula.

FECUNDATION.

TABLE XI.
GENETIC CYCLE OF THE POLYZOA.

PROTOMORPHIC STAGE,
(merged in the following).
The ciliated Germ.

ORTHOMORPHIC STAGE.
Double Polype-embryo.
Polypes.
Ramose Zoophyte—monœcious or diœcious;
with male and female polypes.

GAMOMORPHIC STAGE,
(merged in the foregoing).

Spermary,	Ovary.
Primary Sperm Cell,	Ovum.
Special Sperm Cells and Spermatozoa.	The Germinal vesicle and Macula.

FECUNDATION.

TABLE XII.
GENETIC CYCLE OF THE SALPÆ.

PROTOMORPHIC STAGE,
(merged in the following).
The primordial Germ-mass.
ORTHOMORPHIC STAGE.
The Embryo, transformed into

The solitary *Salpa*.
GAMOMORPHIC STAGE.
Catenated *Salpæ*.

Spermary,	Ovary.
Primary Sperm Cell,	Ovum.
Special Sperm Cells and Spermatozoa.	Germinal Vesicle and Macula.

FECUNDATION.

TABLE XIII.
GENETIC CYCLE OF THE TREMATODA.

PROTOMORPHIC STAGE.
The Infusorial Germ.

The primary Gregariniform Zooid (*Redia or Sporocyst.*)

[The secondary Gregariniform Zooid].

ORTHOMORPHIC STAGE.
Cercaria, penetrating into the tissues of animals.
The same encysted,
And then metamorphosed into a *Distoma*.

GAMOMORPHIC STAGE,
(distinguished from the former by the acquisition of sexual organs).

Spermary,	Ovary.
Primary Sperm Cell,	Ovum.
Special Sperm Cells and Spermatozoa.	Germinal Vesicle and Macula.

FECUNDATION.

TABLE XIV.
GENETIC CYCLE OF THE CESTOIDEA.

PROTOMORPHIC STAGE.

The six-hooked Germ, penetrating into the tissues and converted into a cyst, which becomes
The *Cysticercus*, or the primary *Echinococcus-vesicle*.
[Secondary and subsequent *Echinococci*].

ORTHOMORPHIC STAGE.
Tænia-head *(Scolex)*.

GAMOMORPHIC STAGE.
Cucurbitiform Proglottides.

Spermary,	Ovary.
Primary Sperm Cell,	Ovum.
Special Sperm Cells and Spermatozoa.	The Germinal Vesicle and Macula.

FECUNDATION.

TABLE XV.
GENETIC CYCLE OF ANNELIDA.

PROTOMORPHIC STAGE.
The ciliated Germ, which is converted into
The Head-segment.

ORTHOMORPHIC PHASE.
Segments of the Body of the Annelidan.

Caudal Zooids [of *Syllis, &c.*]

Male and	Female.
Spermary,	Ovary.
Primary Sperm Cells,	Ovum.
Special Sperm Cells and Spermatozoa.	Germinal Vesicle and Macula.

FECUNDATION.

TABLE XVI.
GENETIC CYLE OF THE ARTICULATA GENERALLY.

PROTOMORPHIC STAGE,
(merged in the following).
The Germinal Mass.

ORTHOMORPHIC STAGE:
The Embryo,
[passing through the conditions of *Larva* and *Pupa*.]
Imago or perfect form,
Male and   Female.

GAMOMORPHIC STAGE,
(merged in the foregoing).

Spermary,	Ovary.
Primary Sperm Cell,	Ovum.
Special Sperm Cells and Spermatozoa.	Germinal Vesicle and Macula.

FECUNDATION.

TABLE XVII.
GENETIC CYCLE OF THE APHIDES.

PROTOMORPHIC STAGE,
(merged in the following).
The Germinal Mass.

ORTHOMORPHIC STAGE.
The Embryo, extruded from the Egg as a Hexapod Larva.

Repeated pullulations of other like Larvæ.

Larvæ at the close of the series passing into Perfect Insects,
(Male and | Female).

GAMOMORPHIC STAGE,
(merged in the foregoing).

Spermary,	Ovary.
Primary Sperm Cells,	Ovum.
Special Sperm Cells and Spermatozoa.	Germinal Vesicle and Macula.

FECUNDATION.

TABLE XVIII.

GENETIC CYCLE OF THE HIGHER ANIMALS GENERALLY, IN THE NON-ALTERNATING SPECIES

PROTOMORPHIC STAGE,
(merged in the following).
The Germinal Mass.

ORTHOMORPHIC STAGE.
The Embryo, passing into
The Typical Form,

Male and | Female.

GAMOMORPHIC STAGE.
(merged in the foregoing).

Spermary,	Ovary.
Primary Sperm Cell,	Ovum.
Special Sperm Cells and Spermatozoa.	Germinal Vesicle and Macula.

FECUNDATION.

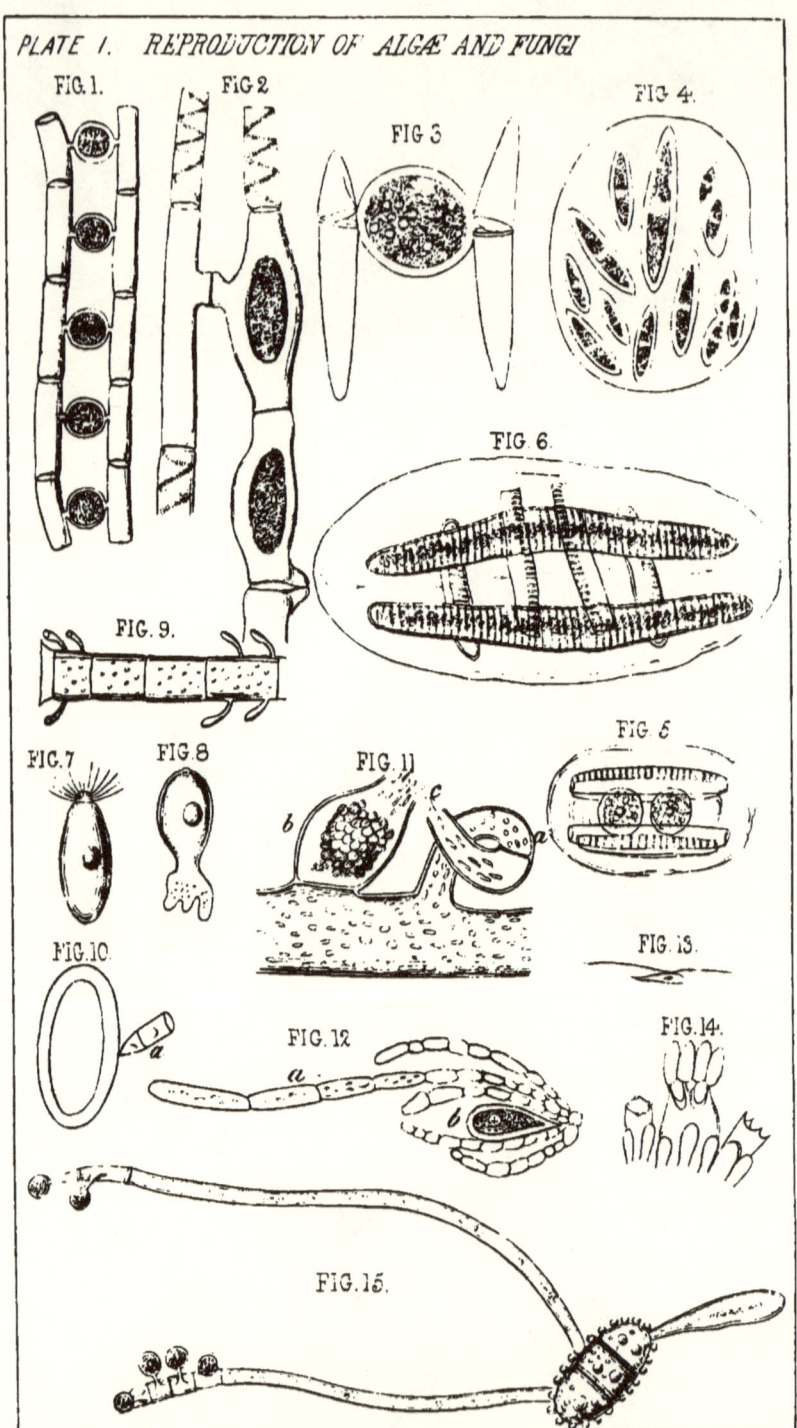

PLATE I. REPRODUCTION OF ALGÆ AND FUNGI

EXPLANATION OF THE PLATES.

PLATE I.—REPRODUCTION IN ALGÆ AND FUNGI.

Figs. 1 and 2.—Filamentous Algæ in conjugation, (p. 23) after Smith. In Fig. 1, *(Mesocarpus,)* the endochrome of both the conjugating cells is evacuated into the connecting tube formed by the fusion of their papillary outgrowths, and the spore is matured in the same place ; in Fig. 2, only one of the conjugating cells evacuates its contents, which are received into the other, where the spore is formed from the fusion of the two endochromes. In this figure conjugation is shown between the cells of distinct filaments, as well as between two adjoining cells of the same filament—a circumstance showing, as Mr. Smith remarks (Trans. Microsc. Soc., II. 70), that the importance of variations in this respect, as distinctive characters, has been over-rated by some Naturalists.

Fig. 3 (from Ralfs) shows the process of conjugation in the Desmidieæ *(Closterium acerosum)*. The evacuation of the cells takes place by the separation of their valves, and the spore is formed, as in *Mesocarpus*, midway between the cells.

Fig. 4 (also from Ralfs) represents a process of endogenous multiplication in the germination of the spore of *Closterium* (p. 27).

Figs. 5 and 6 (*Eunotia turgida*, from Thwaites), illustrate the process of conjugation in the Diatomaceæ, in which the effused endochromes are resolved either at once, or at a very early period, into two spore-masses, which are ultimately transformed into frustules of a much larger size than their parents. The genetic cycle is supposed to be completed by their breaking up into frustules, like those originally concerned in the process of conjugation (p. 28).

Fig. 7.—Ciliated zoospore or motile gemma of *Œdogonium vesicatum*.

Fig. 8.—The same in germination. Having lost its crown of cilia, it has begun to emit radical filaments from its base (p. 32). (Both from Thuret).

Figs 9 and 10 (from Thwaites and Berkeley), illustrate reproduction by androspores, or minute phytoids, formed by the germination of a kind of zoospores, and consisting of a pedicle and two antheridial cells, containing each a single antherozoid. In Fig 9, a filament of *Œdogonium* is shown, with several androspores, and in Fig. 10, a similar body is seen attached to a spore cell of *Bulbochæte crassa*—(p. 33).

Fig. 11.—Reproduction in *Vaucheria* (from Pringsheim); *a*, the "hornlet," or antheridium; *b*, the sporangium, both at first diverticula of the general cavity of the filament, but afterward shut off so as to form distinct cells; *c*, micropyle or terminal pore in the sporangium, by which the antherozoids gain access, when liberated by the dehiscence of the point of the hornlet. The germinal body is at first covered only by a layer of mucus, penetrable by the spermatic particles, but after impregnation, this becomes consolidated into the proper coat of the spore—(p. 34).

Fig. 12.—Reproductive organs of *Sporochnus Adriaticus* (from Kutzing), illustrating the arrangement in the higher Algæ.—*a*, antheridial filament; *b*, sporangium. These parts are more generally separated, and are frequently situated on different plants—(p. 36).

Fig. 13.—Antherozoid of *Fucus vesiculosus* (from Thuret)—(p. 32.)

Fig. 14. (from Berkeley).—Part of the hymenium of *Agaricus velutinus*, illustrating the acrogenous fructification of the higher Fungi. *a*, basidium, with four stalked spores on its summit—(p. 42-45).

Fig. 15. (after Currey).—Compound spore of *Phragmidium bulbosum* in germination, showing the production of threads of mycelium from each of the compartments into which it has become divided, and the secondary spores formed on these filaments in the acrogenous way—(p. 44).

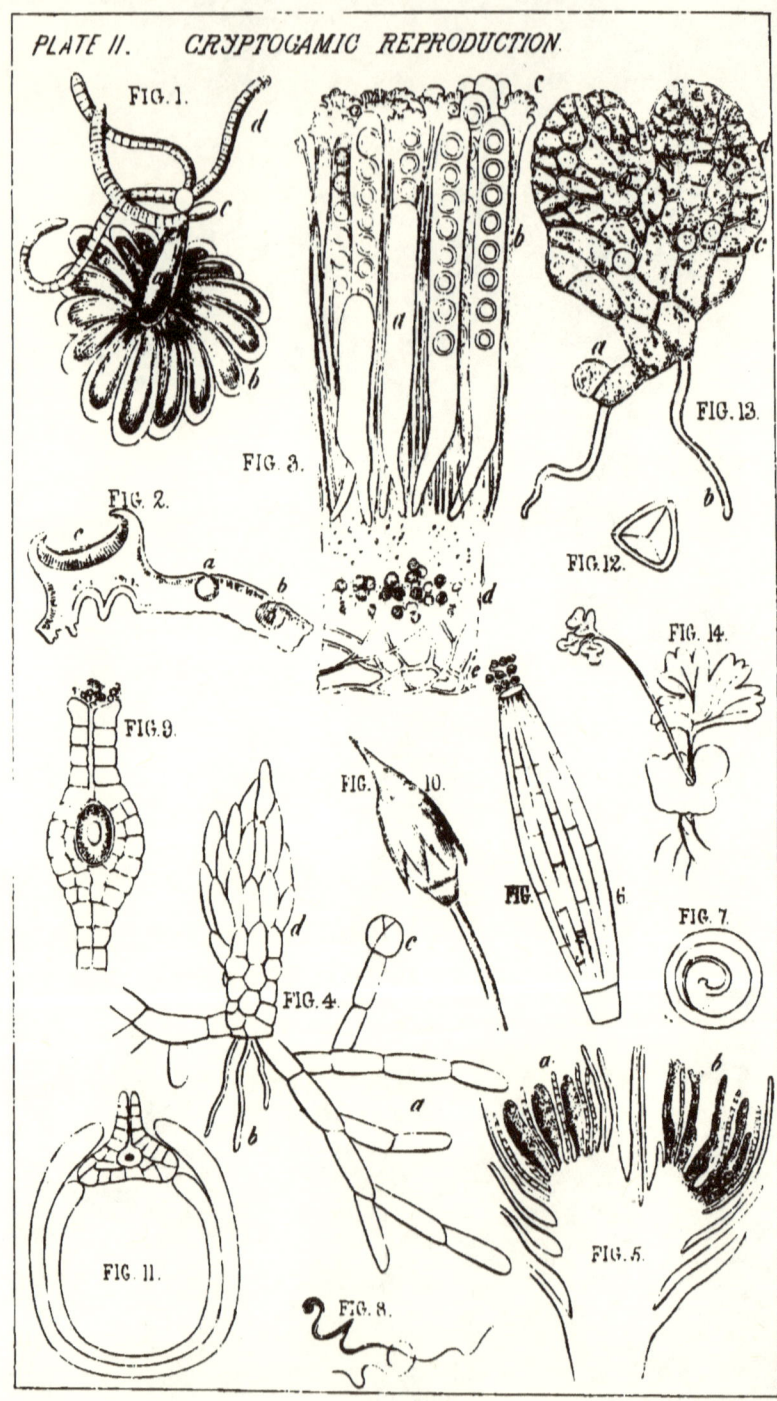

PLATE II.—REPRODUCTION IN THE OTHER CRYPTOGAMIC ORDERS.

Fig. 1.—One of the valves of the globule of *Chara fragilis* (from Berkeley). *b*, cells forming the valve, radiating from the summit of the supporting cell or pedicle, *a* ; *c*, bases of other pedicles springing from the summit of the central stalk of the globule ; *d*, one of the tuft of articulated tubules, which spring from the same point, each of the articulations developing a bi-cilated spermatic filament.—(p. 39).

Fig. 2.—Section of a part of the frond of *Sticta pulmonaria* (after Tulasne), showing spermagonia *a*, *b*—the latter empty—and an apothecium *c*.

Fig. 3.—Section of part of an apothecium of *Lecidea lugubris*, (after Lindsay,) showing immature asci at *a* ; and asci with eight mature spores at *b*; sterile filaments or paraphyses at *c*; the subjacent layer of gonidia at *d*; and the hypothallus or stratum of mycelial filaments at *e*—(p. 48).

Fig. 4.—Germination of a Moss (*Funaria hygrometrica.*) *a* protonemic filaments, the first outgrowths from the spore ; *c*, a nascent moss-bud ; *d*, another bud which has begun to develope a leafy axis, and to emit below radical filaments *b*.—(from Schimper).

Fig. 5.—Perpendicular section of the monœcious fructification of a Moss *(Phascum cuspidatum* from Hofmeister), showing antheridia at *a*, and archegonia at *b*, both intermixed with articulated filaments or paraphyses—(p. 50).

Fig. 6.—Antheridium of a Moss, emitting the cellules which contain the motile filaments.

Fig. 7.—The Antherozoid of a Moss coiled up in its cellule. It is a long simple filament, with one extremity enlarged, and the other tapering away to a fine point.

Fig. 8.—An antherozoid of *Pellia* (from Thuret).

Fig. 9.—Archegonium of *Jungermannia bicaricata* (from Hofmeister) showing the central corpuscule and the styloid canal formed of four columns of cells. At its outer extremity are represented antherozoids, free and in cellules.

Fig. 10.—Capsule of *Polytrichum* (from Balfour), showing the veil, which is derived from the upper wall of the archegonium,

calyptra or while the fruit-stalk, capsule, lid, and contained spores, are all developed out of the germinal corpuscle, which occupied its central cavity.

Fig. 11.—Section of the germinating spore of *Pilularia* (after Hofmeister). At the lower part of the figure are seen the nucleus and two coats of the spore. The pyramidal body at the top is the prothallium, a structure of later growth, here represented as protruding through the micropyle. The section shows an archegonial cavity in its interior—(pp. 57, 183).

Fig. 12 and 13 (after Suminski) illustrate the process of germination in Ferns—(p. 52, 183). Fig. 12.—A tetrahedral fern-spore.

Fig. 13.—Prothallium or cellular outgrowth from the spore *(Pteris serrulata)*. *a*, remains of the spore ; *b*, radical filaments; *c*, antheridial cells containing cellules with ciliated filaments in their interior ; *d*, archegonia.

Fig. 14. Shows the connection of a young Fern *(Pteris paleacea)* developed from the archegonial corpuscule, with its prothallial matrix.

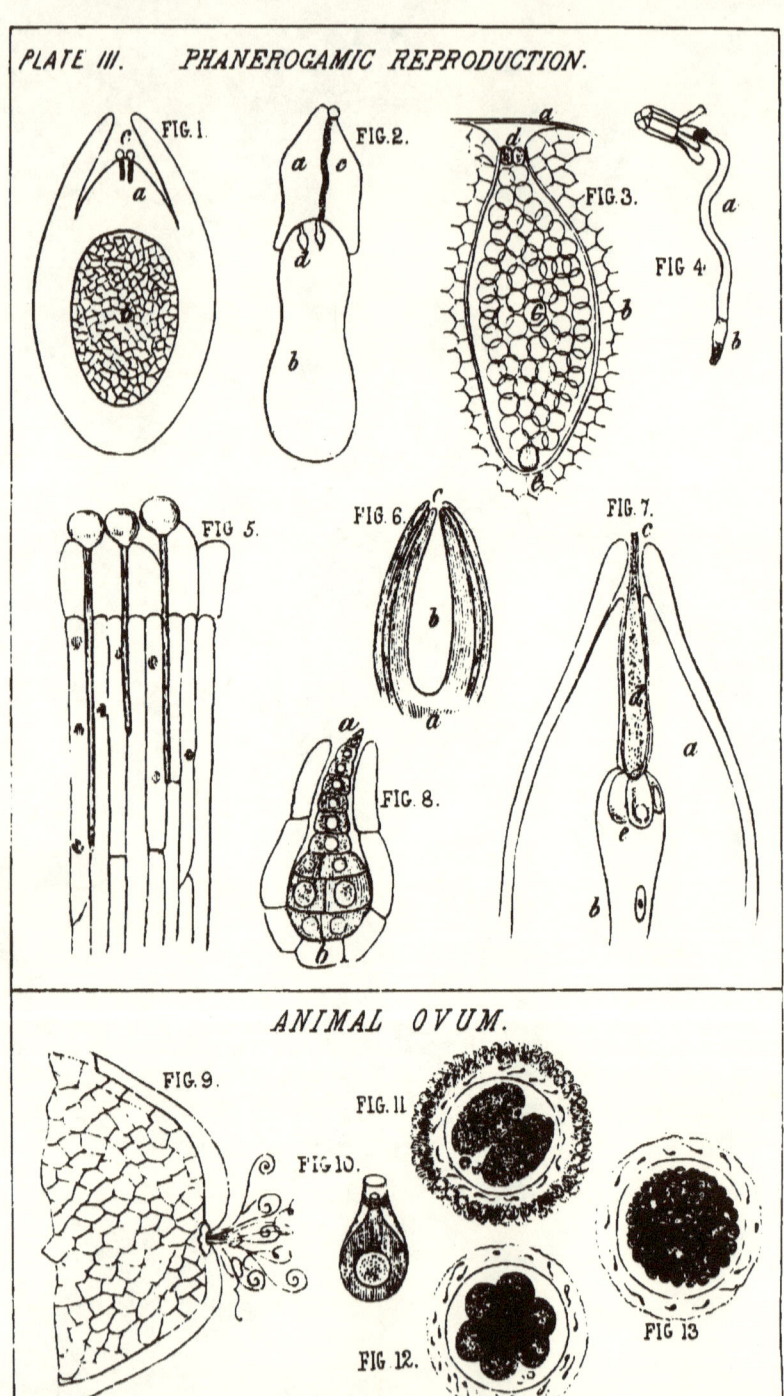

PLATE III.—EMBRYOGENY IN PHANEROGAMIC PLANTS.

Figs. 1-4 (from Hofmeister) illustrate the process in the Coniferæ (pp. 57, 183).

Fig. 1.—A section of the naked ovule of *Pinus sylvestris* in the spring of the year after impregnation. In the micropyle at *c* are seen two pollen grains, with their tubes turned in the direction of the internal cellular mass *b*, going under the name of the *albuminous body*, and corresponding to the prothallium of the higher Cryptogamia.

Fig. 2.—Part of a similar section of the ovule of *Pinus Strobus*, in early summer; *a* upper part of the nucleus, which the pollen-tube, after a long period of inaction, has now nearly traversed, in its way to impinge on the wall of the albuminous body *b*, immediately over one of the clusters of flask-shaped cavities or corpuscula *d*, which have just been formed in that body, and which seem to correspond to the archegonia of the higher Cryptogamia on the one hand, and to the embryo-sacs of the majority of the Phanerogamia on the other.

Fig. 3.—A magnified view of a corpusculum of *Pinus sylvestris*, showing at *e* the germinal particle which is developed into the embryonic structures on impregnation, and, at *d*, two of the four cells which represent the canal of the archegonium—*b* tissue of the nucleus, *c* cellular contents of the corpusculum.

Fig. 4 illustrates the result of impregnation, in the formation of a cluster of four pro-embryos, three of which have been cut away near the base; *b* proper embryo formed by a new process of cell formation at the lower extremity of the suspensor *a*. (*Pinus Strobus*).

Figs. 5-8 illustrate the same process in the Angiospermous Phanerogamia—(p. 61).

Fig. 5 (from Schleiden) shows the emission of tubules by the pollen grains, and their penetration through the conducting cellular tissue of the style.

Fig. 6.—Section of an ovule of *Polygonum divaricatum*, (from Schleiden) showing the seed-coats, the micropyle *c*, and the nucleus *a*, with its large embryo-sac *b*.

Fig. 7.—Ovule of *Œnothera* (from Hofmeister), showing

a pollen tube *d*, passing down from the micropyle *c*, through the tissue of the nucleus *a*, to the embryosac *b*, whose summit is occupied by protoplasmic masses *e*, one of which on impregnation developes the embryo.

Fig 8.—Embryogeny of *Orchis Morio* (from Henfrey), *a* the pro-embryo formed from the upper of the two cells, into which the germinal body divides on impregnation, but, unlike the suspensor of the Coniferæ, growing upwards into a blind filament, which projects through the micropyle ; *b* the true embryo, produced by the sub-division of the lower of the two primary cells.

ANIMAL EMBRYOGENY.

Fig 9.—Part of the egg of *Musca vomitoria* (from Meissner), showing the micropyle occupied by a cluster of spermatozoa—(pp. 69-70).

Fig. 10.—Ovum of *Unio* (from Hossling), in which the aperture of the hollow pedicle serves the purpose of a micropyle—(pp. 69.70).

Figs. 11-13.—Successive stages of the cleavage of the yolk, after impregnation in the Mammalian ovum (from Kirkes)—(p. 72).

PLATE IV.—REPRODUCTION IN POLYPIFERA AND ECHINODERMATA.

PLATE IV.—REPRODUCTION IN POLYPIFERA AND ECHINODERMATA.

Fig. 1.—*Hydra viridis* (from Thomson), showing sexual organs of both kinds and of very simple structure ; *a* spermatic capsule, *b* ovarian cyst containing a single ovum—(pp. 120-124).

Fig. 2.—Reproductive organs of *Campanularia Loveni* (from Steenstrup, after Lovèn). The ovigerous capsule or gonophore is surmounted by two sessile sporosacs or meconidia, *a*, *b*, more complicated in structure, and more approaching to the organization of free medusoids—*a* immature sporosac containing two ova ; *b* sporosac discharging the young in the guise of ciliated infusorians—(p. 123).

Fig. 3.—Another species of *Campanularia*, detaching free medusoids (from Desor)—*a* polype-cell, *b* immature gonophore, *c* gonophore discharging free medusoids *d*.

Fig. 4.—Medusoid of *C. gelatinosa* (from Beneden)—*a* umbrella, *b* manubrium, *c* marginal tentacles.—(p. 122).

Fig. 5.—Infusorian zooid from the ovum of a *Medusa*—(pp. 120-126).

Fig. 6.—Hydraform phase of a Medusa—(*Scyphistoma* of Sars—*Hydra Tuba* of Dalzell.)

Fig. 7.—Hydraform stock detaching from its summit a pile of nascent *Medusæ* (from Dalzell).

Fig. 8.—*Bipinnaria* or precursory zooid of a starfish (from Müller)—(pp. 85,162). At the upper extremity is seen the rudiment of the starfish. The zooid has a mouth and vent of its own, *a*, *b*, but internally its alimentary canal becomes continuous with that of the Echinoderm.

Fig. 9.—*Plutens* or precursory zooid of an *Ophiura*—(from Müller)—at *a* is seen the rudiment of the young Echinoderm, with a distinct mouth of its own in the centre of the disc.

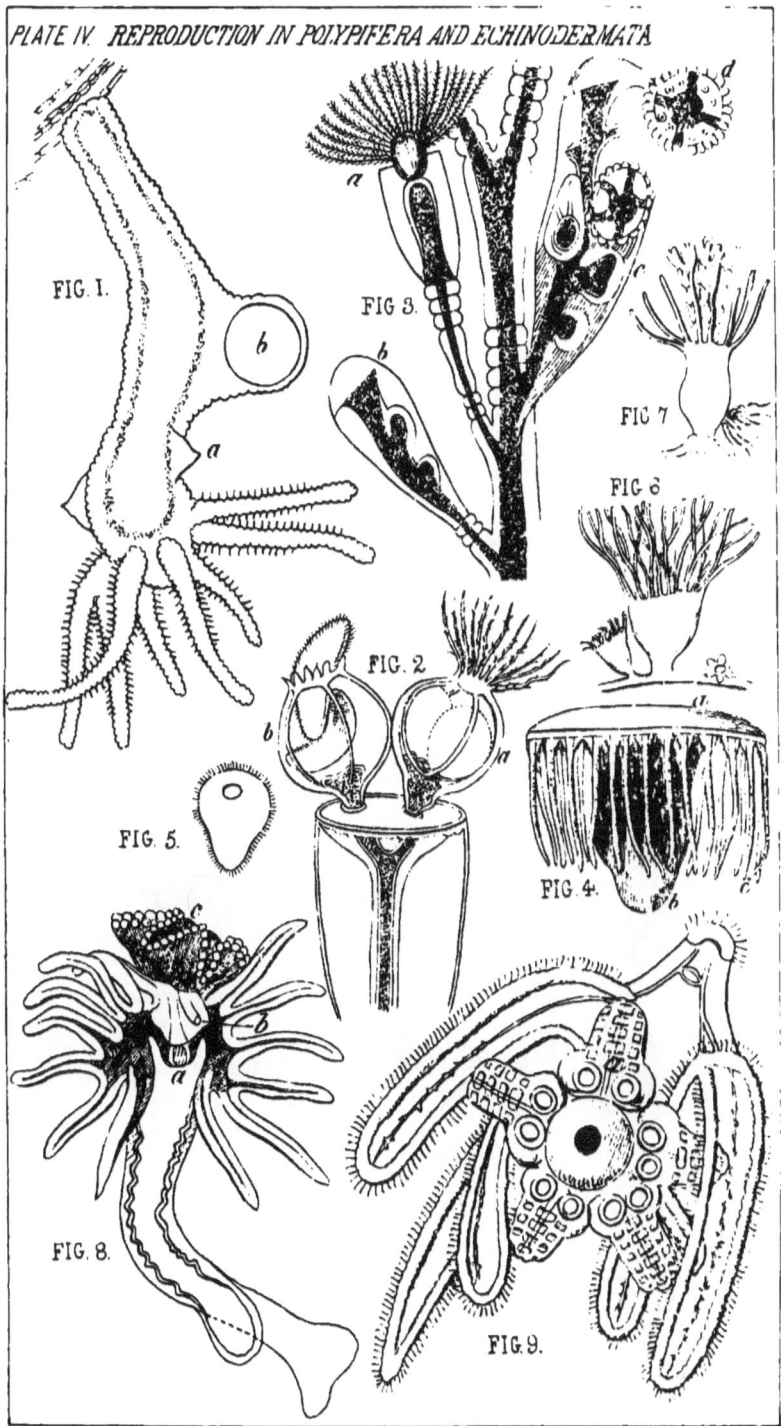
PLATE IV. REPRODUCTION IN POLYPIFERA AND ECHINODERMATA

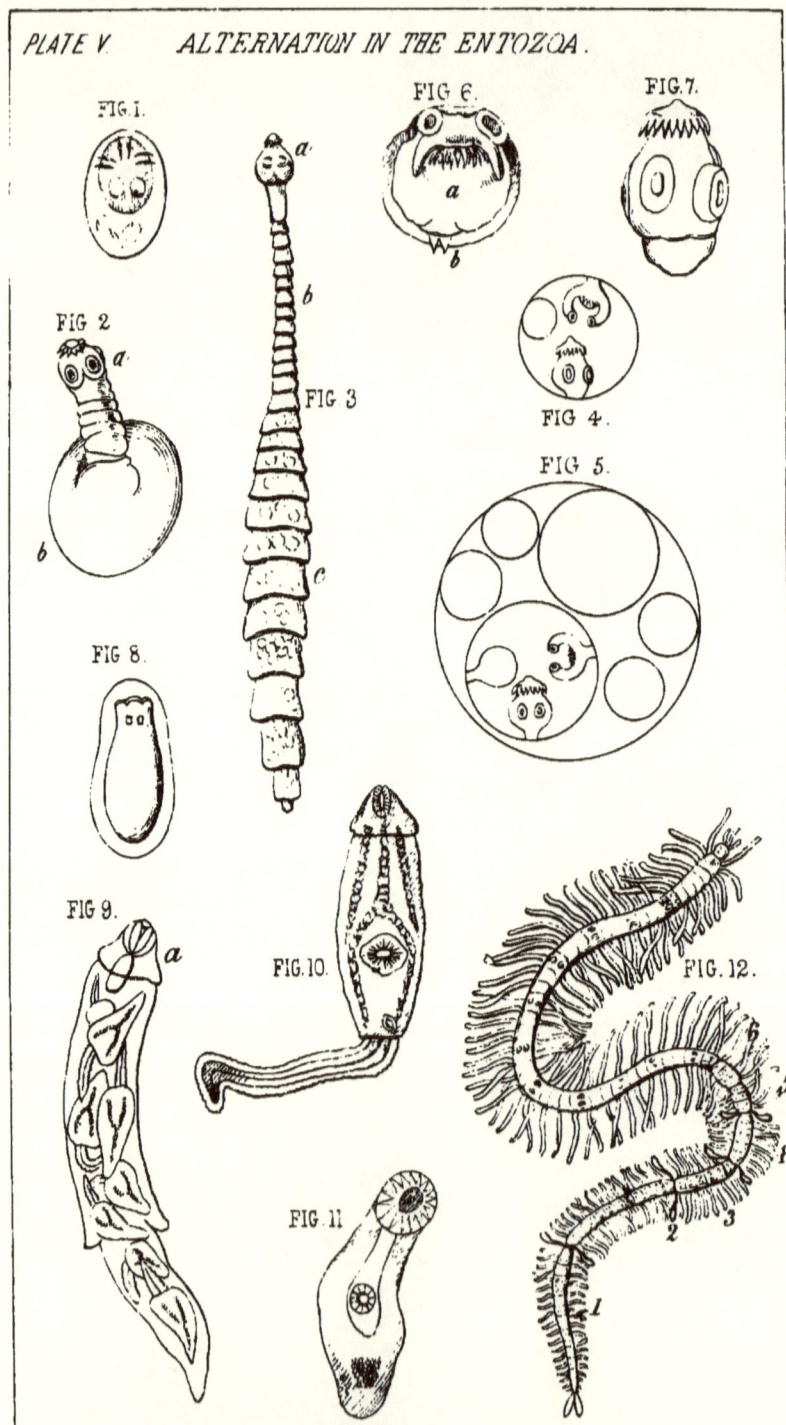

PLATE V.—ALTERNATION IN THE ENTOZOA, &c.

Figs. 1-3 illustrate the successive phases in the development of the Tapeworm—(pp. 96, 131, 161).

Fig. 1.—The egg of a species of Tapeworm, *Tænia Pistillum*, (from Dujardin), with its six-hooked contractile embryo.

Fig. 2.—The *Cysticercus* (from Bremser), now known to be the cystic phase of the *Tænia Solium* ; *a* the head, formed originally within the cyst *b*, and then extruded by the evagination of its hollow pedicle or neck. In the farther development within the alimentary canal, the cyst is thrown off, and the neck extended, by the gemmation and multiplication of its joints, into the long body of the Tapeworm.

Fig. 3.—Ultimate development of the *Tænia Pistillum* of the Shrew (from Dujardin)—*a*, the head ; *b*, the neck, formed of undeveloped segments ; *c*, the body, formed of segments which have acquired their full size, and become filled with ova.

Figs. 4-7 illustrate the development of the *Echinococcus* or Hydatid—(p. 131).

Fig. 4.—A diagram of the simple Hydatid cyst—(*Echinococcus scolicipariens*)—the first result of the transformation of the contractile vesicle discharged from the egg of a minute species of *Tænia*. Derivative Tænia-buds are represented in the course of development from the interior of the cyst—(p. 131).

Fig. 5.—A diagram of the compound Hydatid (*Echinococcus altricipariens*), resulting from the intermediate gemmation of secondary cysts, prior to the formation of proper Tænia-buds.

Fig. 6.—A Tænia-bud, with its denticulated head *a* in process of development, on an invagination of the wall of the cystic portion *b* ; *c* pedicle of attachment to the parent cyst.

Fig. 7.— A fully-formed Tænia-bud, with its head everted (both from Wilson).

Figs. 8-11 (mostly from Steenstrup) illustrate the successive forms in the alternation of the Trematoda—(pp. 95, 112).

Fig. 8.—Egg of *Monostomum mutabile* (from Siebold) with its infusorian embryo.

Fig. 9.—Redia or intermediate form of *Distoma*, derived from the infusorian, and forming in its own interior an agamic brood of

Cercariæ: a œsophogeal cæcum, or rudiment of an alimentary canal. This is really the second of the redial forms figured by Steenstrup.

Fig 10.—*Cercaria*, or larva form of the *Distoma*, being one of the brood figured in the last.

Fig. 11.—Ultimate form of *Distoma*, produced by a transformation of the last during its stage of encystment.

Fig. 12.—*Myrianida fasciata* from M. Edwards—showing the formation of caudal zooids, as described at page 101.

PLATE VI. ALTERNATION IN MOLLUSCA AND ARTICULATA.

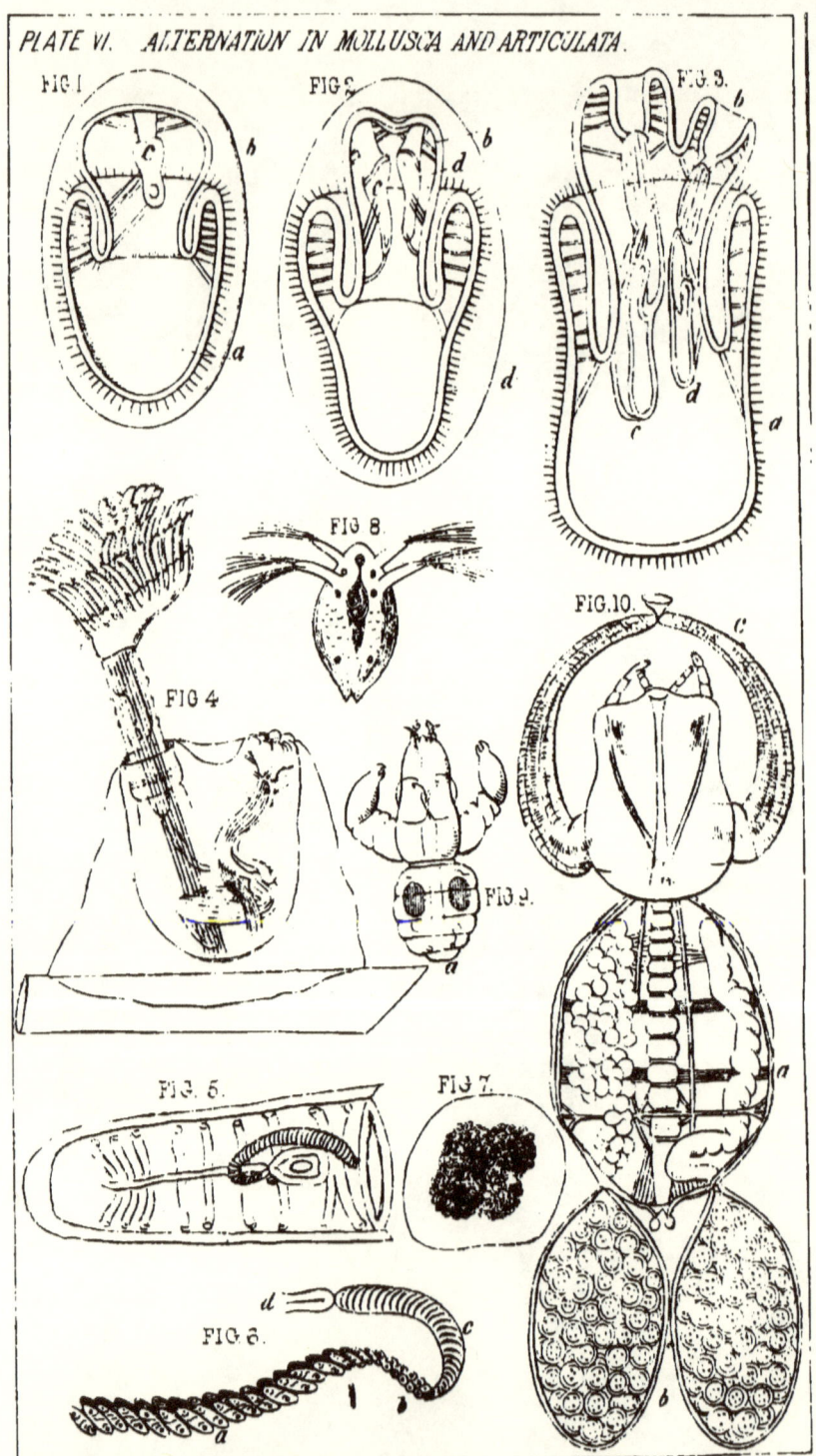

PLATE VI.—ALTERNATION IN MOLLUSCA AND ARTICULATA.

Figs. 1-4. (*from Allman*) illustrate the embryogeny of the Polyzoa (p.p. 88, 160.)

Fig. 1.—Ovum of *Alcyonella fungosa*, containing an embryo in an early stage, with the rudiment of a single polype; *a*, ciliated coat of the embryo; *b*, protrusible non-ciliated portion; *c*, nascent polype.

Fig. 2.—The same, farther advanced with the rudiments of two polypes *c. d.*

Fig 3.—The same, still farther advanced, with both polypes fully formed.

Fig. 4.—A polypidom of *Lophopus crystallinus*, containing two polypes. This is altered from Professor Allman's figure, so far as to show one of the polypes exserted.

Figs. 5 and 6 (from Sars) illustrate the alternation of the *Salpæ*—(pp. 90, 140).

Fig. 5.—A "solitary" *Salpa*, possessing no organs of sex, but budding off from a sort of internal stolon, connected with the nucleus *a*, a pile of sexual zooids *b*.

Fig. 6.—Pile of zooids more enlarged; *a*, the first formed portion of the pile, composed of well developed zooids, and now ready to break off as a Salpa-chain; *b*, middle portion of less advanced zooids; *c*, last formed portion, in which the segments have not yet acquired proper organization; *d*, point of origin from the nucleus.

Fig. 7.—An Ovum (of Eolis?), which, after undergoing the usual segmentation, has divided into four lobes, as if for the development of as many distinct embryos—(p. 160 n.)

Figs. 8-10 (from Nordmann) illustrate the metamorphosis of the Lernean Crustacea—(p. 172).

Fig. 8.—Larva of *Achtheres*, a parasite on the Perch.

Figs. 9 and 10.—The fully-developed male and female of the same—the latter acquiring a much greater proportionate size. Here the abdominal region, *a*, developed in the course of the metamorphosis, may be considered as a gamomorphic structure for the evolution of the organs of reproduction. In the female, besides the large external egg-sacs, *b*, the dissection in fig. 10 shows the cavity of the abdomen to be mainly occupied by the voluminous internal ovaries.

INDEX.

	Page.
Acineta-forms	80
Achlya, Reproduction	34
Algoid or fungoid nature	47
Adventitious organization of zooids	175
Æcidium, Reproduction	43
Affinities in relation to Alternation	219
Agamic eggs	194, 198
Agamozooid	136, 247
Albuminous body	58, 183
Alcippe, Defective Organization	176
Algæ, Reproduction	20, 37
Varieties of Gemmæ	222
Allman, Prof., Organs of Hydrozoa	122, 124, 148
Development of the Polyzoa	160
Reproduction of the Polyzoa	170
Allophytoids and Allozooids	152
Alternation of Gemmations	221
Alternation of Generations	11, 212
Varieties of	12, 19, 109, 136
Formulæ of	216
Contrast in Plants and Animals	107
Concurrence of varieties in the same species	131, 137, 219
Representation in the higher species	109, 144, 157
Simulation of	221
Casual	221

	Page
Ambiguities of Nomenclature	126 n, 132, 153 n, 227
Analogies in Vegetable Reproduction, Tables	258, 264
Androspore	33
Angiospermous Phanerogamia, Reproduction	61
Animals, Survey of Reproduction	68
Tabular views	254, 265, 271
Annelida, Reproduction	99
Alternation	137
Tabular view	269
Antheridium	33, 36, 50
Antherozoids	32, 50
Access of	34, 63 n
Ants, Metamorphosis	105
Neuters	168
Aphis, Reproduction	15, 133, 149
Nature of germs	191
Tabular view	270
Apis, see Bee.	
Apothecium of Lichen	48
Apus, Absence of Males	206
Archegonium	21
Of Moss	50
Homologue of the Embryo-sac	181
Argonauta, Hectocotylus	179 n
Articulata, Reproduction	102
Tabular view	270
Ascaris, Impregnation	69
Germinal Development	72

INDEX.

	Page.		Page.
Ascidia, Metamorphosis	91	*Bothriocephalus*, Development	162
Ascomycetes, Reproduction	42	Botryllida, Reproduction	141
Ascospore	42	*Bovista*, Fructification	45
Ascus of Fungus and Lichen	42	Braun, Prof., Conjugation	26, 27, 35
Asexual Reproduction	9	Pullulation	149
Asplanchna, Defective Organization	176	Vegetable Parthenogenesis	199
		Brightwell, Organization of *Asplanchna*	176
Asteracanthion, Reproduction	84		
Asteridæ	85	Bryozoa, Bryozoaria, see Polyzoa.	
Atavism	222	*Bulbochæte*, Reproduction	33, 34
Auricularia	85	Bulbs, Deciduous	154
Baird, Reproduction of Entomostraca	195	Burnett, Dr. W., Heterogony of *Synapta*	92
Balbiani, Reproduction of Infusoria	77	Germs of *Aphis*	191
		Calycophoridæ, Gemmæ, and Medusoids	83, 118, 124
Bark-lice, Reproduction	193		
Barry, Dr. M., Penetration of Spermatozoa	69 n	Carpenter, Prof., Embryogeny of Mollusca	91, 92
Bary, Reproduction of Fungi	46	Metamorphosis of Insects	104
Vegetable Impregnation	62	Gemmation of Polypes	119, 177
Basidium of Fungus	42, 45	Larval Gemmation	136
Batrachia, Penetration of Spermatozoa	70 n	Embryogeny of the Echinodermata	163
Cleavage of Ovum	203	Of *Planaria*	227 n
Bee, Neuters	105, 168	Carter, Reproduction of *Volvox*	28
Drone Ova	196	Of *Œdogonium*	34, 63
Protracted fertility of the Queen	204	On the Tank worm and the Guinea worm	198
Beneden, Prof. Van, Nomenclature	9 n, 132		
		Carus, Germs of *Aphis*	192
Reproduction of Trematoda	95	Casting of Skin and Limbs	223
Of Cestoidea	98	Casual Alternation	221
Berkeley, Rev. M. T., Morphology of *Chara*	38	Caterpillar	105
		Caryophylleus	96
Reproduction of Fungi	40, 46	Cephalopoda, Embryogeny	91
Bipinnaria of Starfish	85	*Ceramium*, Reproduction	36
Birds' Ovum	73	*Cercaria* of Trematoda	95, 113
Birth in relation to other epochs	107	Cestoidea, Development	96
Bisexual arrangement	241	Contrast with Trematoda	153
Bischoff, Penetration of Spermatozoa	69 n	Gemmation of Embryos	161
		„ of Sex Organs	173
Blastostyle	125	Tabular view	269

INDEX.

	Page.		Page.
Chamisso, Alternation of *Salpa*	90	Corpusculum of Coniferæ	58, 183
Characeæ, Reproduction	38	*Cosmarium*, Gemmation of Spores	27
Parthenogenesis	200	Coste, Spermatophores of	
Morphology	39	Crustacea	226 n
Chermes, Reproduction	134 n	Crustacea, Spermatozoa, 68, 177,	
Development of unimpregnated Ova	193		230 n
		Spermatophores	226 n
Cirripedia, Retrograde Metamorphosis	106, 172	Crust of Lichen	49
		Cryptogamia, Reproduction, 21 *et seq.*	
Defective Organization	175	Contrast with Phanerogamia, 65 180	
Complementary Males 176, 246 n		Tabular view	258, 263
Citrus, Polyembryony	184	*Cryptophialus*, Defective Organization	
Claviceps, Development	44		176 n
Closterium, Development	27	Impregnation	236 n
Coccida, Reproduction	136 n	Cupressinæ, Pollen	59
Unimpregnated ova	193	Currey, Reproduction of Fungi	42
Cocoon, intra-ovular	104	Ctenophora	83 n
Cœlenterata, Reproduction	81	Cycle, see Genetic.	
Cœlobogyne, Parthenogenesis	200	*Cyclops*, Reproduction	136
Cœnurus, Development	131, 161	*Cydippe*, Reproduction	83 n
Columella of Medusa	122	*Cylindrospermum*	28
Comatula, Development	87	*Cynips*, Absence of Males	206
Complementary Males	176, 246 n	*Cypris*, Absence of Males	194
Concentric fission	223	*Cysticercus*, Development	161
Conceptacle of Florideæ	36	Cystic Worms, Development 96, 161	
Conclusions, Summary of	212	*Cytæis*, Pullulation	147
Concurrence of different forms of Alternation	219	*Dacrymyces*, Proliferous spores	45
		Dalrymple, Males of Rotifera	176
Confervoideæ, Reproduction	32	Danielssen, Reproduction of Gasteropoda	91
Coniferæ, Reproduction	57, 183		
Analogies with Cryptogamia	186	*Daphnia*, Reproduction	136
		Ephippial ova	194
Coniomycetes, Reproduction	42	Darwin, Male Cirripedes 176, 236,	
Conjugatæ, Reproduction of	25		246 n
Conjugate-spore	27	Daubeny, Prof., Experiments on Spontaneous Generation	3
Conjugation	20, 23		
In Fungi	47	Davis, Parthenogenesis of	
In Infusoria, &c.	79	Egger Moth	208
Table of Modifications	26	Decapoda, Development of Spermatozoa	177
Cordyceps, Cordyliceps, Development	44	Deciduous Gemmæ	151

INDEX.

	Page.
Deecke, Observations on Vegetable Parthenogenesis	200
Defective Organization	175
Dentition	178
Derivation of Organic beings	1
From one or two parents	9
Desmidieæ, Reproduction	24, 31
Desor, Reproduction of Polypifera	99, 126
Detachment, Physiological Import	177
Development, Epochal acts	107
Of Unimpregnated ova	194
Incipient	202
Diatomaceæ, Reproduction	24, 28, 31
Digenesis	9
Diœcious arrangement	247
Diporpa, Conjugation	79 *n*
Distoma, Reproduction	95, 112
Pullulation	147
Maturation of Organs	168
Double Monsters	158
Drone ova, agamic	197
D'Udekem, Development of Infusoria	80
Dualism	42
Duplicity, Malformation	158
Dzierzon, Reproduction of the Bee	196
Earthworm, Penetration of Spermatozoa	70
Egg-sacs	226
Ecdysis	223
Echinaster, Reproduction	84
Echinodermata	84, 163
Tabular view	267
Echinococcus, Development	132
Pullulation	147, 153
Ectozoa	95
Edward's, Milne, Experiments on Spontaneous Generation	3
Gemmation of *Myrianida*	101

	Page.
Egg-cases or sacs	226
Egg-coverings, Rupture	107
Egger Moth, Parthenogenesis	208
Egg and Seed Structures	235
Embryo of Plants	63, 182
Embryogeny, Phenomena of	18
Representative of one form of Alternation	156, 164
Embryonal vesicles of ovule	62
Embryonic gemmation	160
Endochrome	20, 23
Endothalloid germination	183
Endosperm	65, 260
Entochoncha, Germinal vesicle of	72
Entozoa, Impregnation	70
Reproduction	94
Conjugation	79 *n*
Eolis, Embryonic division	160
Ephippial ova	194
Epochal acts in Reproduction	107
Equisetaceæ, Reproduction	55
Ergot, Development	44
Eschricht, Prof., Development of Entozoa	97
Exothalloid germination	183
Explanation of Tables	253, 260
Extrusion of ova	107
Exuviation	223
Fabre, Development of *Scutigera*	102
Farre, Dr. A., Penetration of Spermatozoa	69
Development of the Ovary	245
Ferns, Reproduction	52
Contrast with Mosses	52, 142, 187
Filaria, Origin	7
Reproduction	99
Filippi, Alternation in *Pteromalus*	113

INDEX.

	Page.		Page.
Fission, as contrasted with gemmation	76 n	Germinal elements	10, 231
		Vesicle of Ovule	62
Concentric	223	Vesicle of Ovum	68
Florideæ, Reproduction	32, 36	Germination of Algæ	27, 30
Fluke-worm, Development	114	Of Fungi	44
Focke, Germination of Closterium	27	Giraldès, "Corps Innominé" of	245
		Globule of *Chara*	38
Food-yolk	74	Gonidium	49
Foraminifera	77	Gonophore	125
Formulæ of Genetic Cycle	216	Goodsir, H. D., Organization of Male Cirripedes	176, 230
Frog, Penetration of Spermatozoa	69	Spermatozoa of Crustacea	177, 208
Fucoideæ, Reproduction	32, 36	Gosse, Organization of Asplanchna	176
Fucus, Impregnation	37		
Incipient Germination of Unimpregnated Spores	202	Gradation of Sex Organs into Zooids	173
Fungi, Reproduction	40	Greene, Prof., Protozoa	76
Germination	44	*Gregarina*, Conjugation	79
Conjugation	47	Griffiths, Conjugation of Diatomaceæ	28
Furcation of Embryo	159		
Gamomorphic Stage	14, 239	Guinea-worm, Origin	98
Alternation	111, 117	Gymnosperms, see Coniferæ.	
Pullulation	147	Hair, Life of	178
Gartner's Canal	244	Hectocotylus	179 n, 225 n
Gasteromycetes, Reproduction	42, 45	Helminths, Reproduction	94, 112, 131, 153, 161
Gemmæ, adherent and deciduous	157	Tabular views	268
Relation to Ova	190	Henfrey, Prof., Germination of *Spirogyra*	27
Bearing on Alternation	200	Polyembryony	184
Gemmation	9	Hepaticæ, Reproduction	49
Relation to Fission	76 n	Hermaphroditism, normal	241
In the higher species	167	In Mollusca	94
Alternation of	221	In Helminths	94
Table of interpolation of	256	In Annelida	101
Generation, Spontaneous	2	In Articulata	106
See Alternation.		Abnormal	242
Genetic Cycle, Formulæ	216, 218	Functional	246
Relation to affinity	219	*Hermella*, Incipient development of Unimpregnated Ova	202
Tabular views	254, et seq.		
Geophilus, Spermatophore	226 n		

INDEX

	Page.
Heterocyst	29
Heterogenesis	9
Heterosporous, Cryptogams	186
Hive Bee, Reproduction	196
Hofmeister, Germination of Cosmarium	27
Reproduction of Mosses and Ferns	52
Reproduction of Coniferæ	57
Holothuridæ, Reproduction	85
Homogenesis	9
Homoiosporous, Cryptogams	186
Homology of the reproductive structures in general	229, 248
Special do.	237, 244
Honey Bee, Reproduction	196
Huxley, Prof., Ovum of *Pyrosoma*	72 n
Spermatozoa of *Tethya*	77
Zoological Individuality	110
Medusoids of Polypifera	125, 174
Gonophores of Polypifera	126
Reproduction of *Salpa*	141
Homology of *Medusæ*	147 n
Germs of *Aphis*	192
Hydra, Reproduction	120
Hydra Tuba, Reproduction	120
Hydractinia, Composite Structure	179
Hydrozoa, see Polypifera.	
Hygrocrocis	47
Hymenium of Fungus	49
Hymenomycetes, Reproduction	42, 45
Hyphomycetes, Reproduction	41, 46
Hypothallus of Lichen	49
Hypothesis, use in Science	17
Ibla, Defective Organization	176
Identity of Ova and Gemmæ	205
Imago of Insect	103

	Page.
Impregnation, Modifications of 70 n,	246 n
Theory	204, 231
Incipient development of Unimpregnated Ova	202
Incompatibility of Ova and Gemmæ	205
Individuality	109, 165
Inessential Sprouts	155
Infusoria	77
Ink Mould	47
Insects, Metamorphosis	102
Insemination, Relation to other epochal acts	107
Interpolation of Gemmation, Table of	256
Isoetes, Reproduction	55
Isophytoids	152
Isozooids	152
Itzigsohn, Development of *Oscillatoria*	29
Jelly-fishes, Reproduction	117
Jenner, Germination of *Closterium*	27
Analogy of Ferns and Mosses	53, 143, 183
Jungermannia, Reproduction	49
Fructification	188 n
Karsten, Conjugation of Siphoneæ	35
Keber, Penetration of Spermatozoa	69 n
King's yellow worms	115
Knox, Dr., Development of Sex	242, 245
Kobelt, Wolffian Bodies	244
Koren on Development of Gasteropoda	91
Küchenmeister, Origin of Parasites	8, 99
Determination of drone ova	197

INDEX.

Lachmann, Reproduction of Infusoria - 80
Lankester, Nomenclature of Gemmæ - 152
Larvæ - 73, 103, 238
Larval Gemmation - 136
Late Development of Sex - 168
Laycock, Prof., Diffusion of Vitality - 179 n
Leech, Multiple Embryos - 226 n
Lereboullet, Double Monsters in Pike - 159
Lerneans, Retrograde Metamorphosis - 106, 172
Léveillé, Development of Ergot 45
Lewis, Gemmation in Annelida - 139 n
Heterogony - 130 n
Leuckart, Development of Trichina - 99 n
Reproduction of Chermes 134 n, 136 n
Reproduction of Aphis 193
Spermatophores of Octopus 225 n
Development of Spermatozoa - 204, 230 n
Homology of the Prostate Vesicle - 244 n
Leydig, Prof., Germs of Aphis 192
Lichens, Reproduction - 48
Life, Somatic and topical 177
Limnadia, Absence of Males 206
Lindsay, Dr. L., Reproduction of Lichens - 48
Links, Variable number in the Genetic Cycle 153
Liquor Seminis - 231
Lizzia, Pullulation - 147
Lolligo, Spermatophores 225 n
Lubbock, Zoological Individuality - 110

Lubbock, Germs of *Aphis* 192
Ova of *Daphnia* - 194
Lucernariadæ, Reproduction 83, 120, 127
Lumbricus, Multiple embryos 69 n
Lycopodiaceæ, Reproduction 55
Macrospore - 55
Maggot - 104
Males, Complementary 176, 246
Absence of - 206
Malformation of duplicity 158
Mammalia, Penetration of Spermatozoa - 70 n
Reproduction - 107
Mangrove, Germination 64
Manubrium - 122
Marchantia, Fructification 188
Marsileaceæ, Reproduction 56
Marsupium - 122
Maturation of Sex - 15, 167
Meconidium - 123
Medusæ, Origin and Relations 117
Direct development 129 n
Pullulation - 117
Morphology - 126, 117 n
Medusoids - 83, 117
Meissner, Penetration of Spermatozoa - 69 n
Meloseira, Conjugation 27
Mesocarpus, Conjugation 26, 27
Metagenesis - 109
Metamorphosis - 73, 102, 223
Microgonidium - 33
Micropyle of Algæ - 32
Of Rhizocarpeæ - 57
Of Phanerogamia - 58
Of the Animal Ovum 70
General Homology - 234
Microspore - 55
Mildew - 46
Mollusca, Reproduction 91

292   INDEX.

	Page		Page
Mollusca, Germinal Segmentation	159	Nomenclature, Ambiguities	126 n, 132, 153 n, 163 n, 227
Monœcious arrangement	247	Non-sexual Reproduction	9
Monogenesis	9	Nostochineæ, Reproduction	31
Monsters, Double	158	Nucules of *Chara*	38, 39
Monostomum, Development	113 n	Octopus, Spermatophores	225 n
Morphology of *Chara*	39	*Œginopsis*, Direct development	130 n
Of Medusæ	126, 147 n	*Œdogonium*, Reproduction	33, 34
Of the Uterus	244	Access of Antherozoids	63 n
Mosses, Reproduction	49	Ophiuridæ, Reproduction	85
Analogy with that of Ferns	53, 142	Organic Bodies, Origin	2
		Organization, Adventitious	175
Mougeottia, Conjugation	26	Origin of Organisms	1
Moulds	41, 46	Orthomorphic Stage	14, 133, 238
Mucor, Development	47	Pullulation in	149
Mulberry body	157	Representation in the higher species	166
Munter, Reproduction of Ferns	57 n, 186	Oscillatorieæ, Reproduction	29, 31
Müller, Johan, Embryogeny of *Entochoncha*	72	Ova, in relation to Gemmæ	190, *et seq.*
		Bearing on Alternation	208
Development of Echinodermata	84	Development, without impregnation	191, 199
Otto, Fission of Annelida	101	Ovary, Development	245
Mullerian Cord	244	Ovigeous, Capsules of Polypifera	123
Murray, Development of Insects	105		
Mycelium of Fungi	40, 49	Ovisac	68
Myrianida, Fission	101, 138	Ovules of Coniferæ	58, 186
Myriapoda, Reproduction	102	Of Phanerogamia	62, 181
Nais, Fission	101	Ovum	68
Nautilus, Spermatophores	225 n	Of Bird	73
Needham, Filaments of	225	Of *Pyrosoma*	73
Nelson, Dr., Penetration of Spermatozoa	69 n	Owen, Prof., Development of *Distoma*	113
Nematoidea, Reproduction	70, 72	Relations of *Medusæ*	121
Nereis, Fission	101	Reproduction of *Aphis*	152
Neuters	161, 241	Identity of Ova and Gemmæ	205
Newport, Penetration of Spermatozoa	69 n	Paget, Life of hairs, &c.	178
		Palmelleæ, Conjugation	23
Germinal Vesicle of Batrachia	71	Germination	27, 28
		Palmoglea, Conjugation	26
Segmentation of Ovum	203, 205	*Paramœcium*, Reproduction	77

	Page.		Page.
Paraphyses of Lichens	48	Polypifera, Pullulation	147
Parasites, Origin	7	Polypidom, Composite nature	
(See Cestoidea, Nematoidea, Trematoda, &c.)		of	110, 119
		Polyzoa, Reproduction	87
Parthenogenesis (True)	194	Tabular view	267
In Plants	199	Embryonic gemmation	160
Pasteur, Spontaneous generation	4	Gamomorphic do.	170
		Pouchet, Experiments on Spontaneous Generation	3
Pectinibranchiata, Reproduction	92		
Pelagia, Direct development	130 n	Proglottis	132
Pelodytes, Hermaphroditism	94 n	Prostate Gland, Homology	245
Penetration of Spermatozoa	70	Prostate Vesicle	244
Of Antherozoids	32, 34, 63	Prothallium	52, 182
Pentacrinus, Development	87	Protomorphic Stage	14, 237
Perichœtium of Moss	50	Alternation	111
Periodicity of Sex	167	Pullulation	147
Phanerogamia, Reproduction	29, 57	Representation	156
Analogies with Crytogamia	60, 65, 180	Protonema of Moss	50, 154, 188 n
		Protophyta, Reproduction	20
Tabular view	262	Tabular view	31
Physalia, Medusoids	125	Protozoa, Distinction from Protophyta	75
Physical Origin of Organic beings	2		
Physosphoridæ, Reproduction	83, 118, 124	Reproduction	76
		Protracted fertility of Queen bee	204
Phytoids	10, 110	Pringsheim, Reproduction of Algæ	27, et seq.
Phytozoa, Phytozoaria, see Antherozoids.			
		Pteromalus, Supposed Alternation	136
Pilularia, Reproduction	186		
Pike, Duplex embryos	159	*Puccinia*, Development	42
Pistillidium	51	Pullulation,	146
Placenta of *Salpa*	90	Formula	218
Plants, see Vegetables.		Pupæ	105
Planaria, Multiple of embryos	227	Pycnidium of Lichen	48
Pluteus of *Echinus*	85	*Pyrosoma*, Development of Ovum	72
Pollen	22	Quatrefages, Fission of *Syllis*	101
Polyembryony	184	Incipient development of unimpregnated Ova	202
Polyphemus, Absence of Males	206		
Polypifera, Reproduction	81, 117	Rabbit, Penetration of Spermatozoa	69 n
Tabular view	266		
Gradation of organs into zooids	124, 174	Radlkofer, Reproduction of Fungi	43 n, 45 n

	Page.
Radlkofer, Genetic Cycle of Cryptogamia	53, 143 n
Vegetable Parthenogenesis	200
Ralfs, Gemmation of *Closterium*	27
Redia of Trematoda	113
Regel, Denial of True Parthenogenesis in Plants	200 n
Regional Life	177
Reissek, Germination of Pollen	203 n
Reid, Gemmation of *Hydra Tuba*	120
Representation of Alternation in the higher species	15
Protomorphic	156
Gamomorphic	166
Reproduction, Sexual and Non-sexual	9
In the Vegetable Kingdom	17
In the Animal Kingdom	68
Homologies	229
Resting-spores	28
Resting-periods, Table of	257
Rhizocarpeæ, Reproduction	56
Robin, Antherozoids of *Ulva*	29
Defective Organization of Males	175
Rosenmuller, Development of the Ovary	245
Rotifera, Ova	195
Rye, Spurred (Ergot), Development	44
Salix, Change of Sex	207 n
Salpa, Reproduction	90
Alternation	140
Tabular view	268
Sarsia, Pullulation	147
Sanderson, Dr., Genetic Cycle of Mosses and Ferns	53, 57
Analogies of Coniferæ	186
Table (modified)	258
Saprolegnia, Development	35

	Page.
Scalpellum, Defective Organization	176
Schacht, Development of Pollen-tube	58, 62
Schleiden, Vegetable Impregnation	62
Schneider, Hermaphroditism of *Pelodytes*	94
Schultze, Experiments on Spontaneous Generation	2
Development of Nemertini	99
Scolex	132
Scutigera, Development	102
Scyphistoma, Development	120
Seeds and eggs, Analogies	235
Segmentation of Ovum	72
Selaginella, Reproduction	55, 186
Seminal Vesicles, Homology	245 n
Sex, Late development	168
Periodicity	167
Theory of original conjunction	242
Gradation of Organs into Zooids	173
Ultimate Corpuscules	231
Disguise and Mixture	244
Sexes, Association in Plants	185
Distribution in Species	241, 247
Sexual reproduction	7
Rudimentary manifestation	20
Siebold, Parasitic *Filariæ*	7
Penetration of Spermatozoa	69
Development of *Distoma*	113
Genetic Cycle of Polypifera	128
Development of *Tænia*	161
Parthenogenesis of the Bee and other Insects	196, 206
Spermatophores	226
Silkworm Moth, Parthenogenesis	207 n
Simpson, Prof., on Hermaphroditism	243

INDEX.

	Page
Siphoneæ, Reproduction	34
Simulation of Alternation	221
Smith, J., of Kew, Parthenogenesis of Cælobogyne	199
Smut, Development	42
Somatic and topical life	177
Soredium of Lichen	49
Special homology of reproduction	237
Spermagonium of Lichen	48
Spermatia of Fungi and Lichens	41
Spermatic Element	10
Liquor	231
Ultimate Corpuscules	229
Homologies	235
Theory of Action	203, 206
Spermatophores	225
Spermatozoa	68, 230, 177
Self-development	203
Sphæria, Reproduction	41, 46
Sphæroplea, Reproduction	33
Spinula of *Ascidia*	91
Spirogyra, germination	25
Sponge, Spermatozoa	77
Spontaneous Generation	2
Spores, Conjugate	27
Resting, or winter	28
Of Fungi	41
Of the higher and lower Cryptogamia	187
Sporocarp	56
Sporocyst	113
Sporophore	225
Sporosac	122
Spurred Rye, Development	44
Star-fish do.	86
Statoblast of Polyzoa	222
Steenstrup, Alternation of Generations	109
Development of *Distoma*	112
Hectocotylus-arm of Cuttlefish	225 n
Stein, Metamorphosis of Protozoa	80
Stolon of Tunicata	90
Strobila	126, 132 n
Stylospore of Lichen	48
Suminski, Germination of Ferns	57 n
Summary of Conclusions	212
Survey of Vegetable reproduction	17
Survey of Animal reproduction	68
Suspensor of Embryo	182, 188
Syllis, Caudal gemmation	100, 138, 180 n
Synapta, Parasite of	92
Fission	87
Syngamus, Conjugation	79
Synœcious arrangement	247
Syzygites, Conjugation	48
Table of the Modifications of Conjugation	26
Of the Protophyta	31
Of the Modifications of Impregnation	71
Of Epochal acts	107
Of the Structure of Eggs and Seeds	235
Of the Homology of Reproduction	236
Of the Protomorphic, Orthomorphic, and Gamomorphic Stages	237, 240
Of Sexual Arrangement	248
Of the Genetic Cycle	254, *et seq.*
Explanations	253, 260
Tænia, Origin and development	96, 131, 161
Tankworm	98
Tapeworm, see *Tænia*.	
Teeth, Proper Life	178
Terebella, Gemmation	139 n
Termites, Neuters	105, 168
Tethya, Spermatozoa	77
Tetrarhynchus, Development	98

INDEX.

	Page.		Page.
Tetraspores	36	Varieties of Alternation	109
Thaumantias, Pullulation	147	*Vancheria*, Reproduction	34
Theca of Fungus and Lichen	42	Vegetables, Composite Cha-	
Of Moss	50	racter	14, 149
Thecaspore	42	Reproduction	17
Theory, Use in science	17	Alternation	142
Thomson, Prof. A., Nomenclature	9	Tabular View	254, 264
Genetic Cycle of Polypifera	128	Pullulation	149
Double embryos	158	*Velella*	118, 124
Prof. Wyville, Development		Vertebrata, Reproduction	106
of *Comatula*	87	Vesicula Prostatica	244
J. V., Development of		Seminalis	245
Comatula	87	Vinegar Plant	47
Thuret, Antherozoids of Ulva	29	*Viscum*, Polyembryony	184
Reproduction of Nostochineæ	29	Vitelligenous cells of Insects	193
Reproduction of Fucoids	37	Vitelline, Membrane	74
Germination of unimpreg-		Viviparous Plants	154 n, 201
nated spores	202	*Volvox*, Reproduction	28
Thwaites, Germination of Palmella	28	Volvocineæ, do.	31
Gemmæ of Algæ	36	*Vorticella*, do.	79
Topical and Somatic life	177	Vrolik, Double Monsters	158
Trachynema, Direct Develop-		Wagner, Development of Sper-	
ment	130	matozoa	204
Trematoda, Reproduction	95, 112	Weber, Development of Sex	244
Tabular View	268	Weeping-willow, Change of sex	207 n
Analogies in Cestoidea	153	Winter eggs	194
Tremellineæ, Reproduction	45	Winter spores	28
Trichina, Development	99 n	Wolffian Bodies, Development	
Tulasne, Reproduction of Fungi	41, 48	and Relations	244
Vegetable Impregnation	62	Worms, Intestinal	94, 153
Tunicata, Reproduction	89	Annelidan	99
Typical form of Polypifera	120, 128	Wright, Dr., Reproduction of	
Udekem, Metamorphosis of		Polypifera	83, 123
Infusoria	80	Organization	147 n, 180
Ulvaceæ, Reproduction	29, 31	Yolk, Cleavage	71
Unimpregnated ova, Develop-		Supplementary	74
ment	191, 199	Zooids	10, 13, 109
Incipient	202	Gradation into Organs	173
Uredo, Development	42, 43	Adventitious Organization	175
Uterus, Homology	244	Zoophytic Organization	14, 82, 149
Valentine, Furcation of Embryo	159	Zoospores	28, 32, 37
Variability of Pullulation	153	*Zygnema*, Conjugation	25

www.ingramcontent.com/pod-product-compliance
Lightning Source LLC
Chambersburg PA
CBHW030016240426
43672CB00007B/968